Esther Mitterbauer

Heat stress in tomato - a challenge for plant breeding

Esther Mitterbauer

Heat stress in tomato - a challenge for plant breeding

Identification of factors limiting fruit set in tomato (Solanum lycopersicum L.) with the aim of genetic improvement of heat tolerance

Südwestdeutscher Verlag für Hochschulschriften

Impressum/Imprint (nur für Deutschland/ only for Germany)
Bibliografische Information der Deutschen Nationalbibliothek: Die Deutsche Nationalbibliothek verzeichnet diese Publikation in der Deutschen Nationalbibliografie; detaillierte bibliografische Daten sind im Internet über http://dnb.d-nb.de abrufbar.

Alle in diesem Buch genannten Marken und Produktnamen unterliegen warenzeichen-, marken- oder patentrechtlichem Schutz bzw. sind Warenzeichen oder eingetragene Warenzeichen der jeweiligen Inhaber. Die Wiedergabe von Marken, Produktnamen, Gebrauchsnamen, Handelsnamen, Warenbezeichnungen u.s.w. in diesem Werk berechtigt auch ohne besondere Kennzeichnung nicht zu der Annahme, dass solche Namen im Sinne der Warenzeichen- und Markenschutzgesetzgebung als frei zu betrachten wären und daher von jedermann benutzt werden dürften.

Verlag: Südwestdeutscher Verlag für Hochschulschriften Aktiengesellschaft & Co. KG
Dudweiler Landstr. 99, 66123 Saarbrücken, Deutschland
Telefon +49 681 37 20 271-1, Telefax +49 681 37 20 271-0
Email: info@svh-verlag.de
Zugl.: Hannover, Leibniz University, Diss., 2008

Herstellung in Deutschland:
Schaltungsdienst Lange o.H.G., Berlin
Books on Demand GmbH, Norderstedt
Reha GmbH, Saarbrücken
Amazon Distribution GmbH, Leipzig
ISBN: 978-3-8381-1245-9

Imprint (only for USA, GB)
Bibliographic information published by the Deutsche Nationalbibliothek: The Deutsche Nationalbibliothek lists this publication in the Deutsche Nationalbibliografie; detailed bibliographic data are available in the Internet at http://dnb.d-nb.de.

Any brand names and product names mentioned in this book are subject to trademark, brand or patent protection and are trademarks or registered trademarks of their respective holders. The use of brand names, product names, common names, trade names, product descriptions etc. even without a particular marking in this works is in no way to be construed to mean that such names may be regarded as unrestricted in respect of trademark and brand protection legislation and could thus be used by anyone.

Publisher: Südwestdeutscher Verlag für Hochschulschriften Aktiengesellschaft & Co. KG
Dudweiler Landstr. 99, 66123 Saarbrücken, Germany
Phone +49 681 37 20 271-1, Fax +49 681 37 20 271-0
Email: info@svh-verlag.de

Printed in the U.S.A.
Printed in the U.K. by (see last page)
ISBN: 978-3-8381-1245-9

Copyright © 2010 by the author and Südwestdeutscher Verlag für Hochschulschriften Aktiengesellschaft & Co. KG and licensors
All rights reserved. Saarbrücken 2010

Table of contents

List of abbreviatons 3
1   Introduction 5
    1.1   Tomato 5
    1.2   Heat stress 5
    1.3   Heat stress in tomato 6
2   Materials and methods 9
    2.1   Plant material 9
    2.2   Growth conditions 10
    2.3   Greenhouse and climate chamber facilities 11
    2.4   Evaluation of phenotypic traits 12
    2.5   Chemicals 13
    2.6   Histological techniques 14
    2.7   Microscopical techniques 16
    2.8   Flow cytometry 17
    2.9   DNA extraction 17
    2.10   DNA quantification 17
    2.11   AFLP Analyses 18
    2.12   Data analyses 23
3   Results 24
    3.1   Highest applicable temperature for heat stress experiments 24
    3.2   Verification of the integrity of the gynoecium 30
    3.3   Histological investigations of androecia and gynoecia grown under heat stress 31
    3.4   Response of different tomato species and genotypes to heat stress in climate chambers 34
    3.5   Comparisons of different methods to evaluate pollen viability and pollen tube growth (*in vivo* and *in vitro*) 38
    3.6   Pollen storage 45
    3.7   Genetic variability in heat tolerant tomato lines under greenhouse conditions in Thailand 46
    3.8   Introgression lines 57

3.9	Plant response to different greenhouse set-ups	60
3.10	Affirmation of heat stress as reason for reduced plant vitality	74
3.11	Phenotypic evaluation of a segregating F_2 population for mapping QTLs for heat tolerance	77
3.12	Comparison of results obtained by flow cytometry with results achieved by microscopy	86
3.13	Analyses of AFLP markers within the segregating F_2 population	88
3.14	Production of F_3 seed for further investigations	90
4	Discussion	91
4.1	Methods	91
4.2	Effects of heat stress on vegetative growth	95
4.3	Effects of heat stress on generative growth	99
4.4	Genetic variability	106
5	Conclusions	112
6	References	115
Appendices		128

List of abbreviatons

'bref'	black ground mulch combined with NIR reflecting pigment on the roof cover
'btrans'	black ground mulch combined with NIR transmissive roof cover
'wref'	white ground mulch combined with NIR reflecting pigment on the roof cover
'wtrans'	white ground mulch combined with NIR transmissive roof cover
µE	micro Einstein
AFE	ethanol; formaldehyde; acidic acid
AFLP	Amplified fragment length polymorphism
AIT	Asian Institute of Technology
ANOVA	analysis of variance
APS	ammonium persulfate
AVRDC	AVRDC - The World Vegetable Center (The Asian Vegetable Research and Development Center)
B&W	Brewbaker and Kwack pollen germination medium
bp	basepair(s)
CTAB	cetyl trimethylammonium bromide
cw	calendar week
DI	deionized water
DNA	deoxyribonucleic acid
dNTP	deoxynucleotide triphosphate
EDTA	ethylenediamine tetraacetic acid
F_1	filial 1, the first filial generation
F_2	filial 2, the second filial generation
FDA	fluorescein diacetate
FAP	greenhouse cooled by a fan and pad cooling system

GC	greenhouse cabinet
GH	greenhouse
IL	introgression line
LSD	least significant difference
MAS	marker assisted selection
MTT	tetrazolium bromide (3-(4,5-Dimethylthiazolyl-2)-2,5-diphenyl-2H-tetrazoliumbromid)
NIR	near infrared
PAS	periodic acid-Schiff stain
PCR	polymerase chain reaction
PE	polyethylene
PS	phenosafranine
QTL	quantitative trait locus
RH	relative humidity
rpm	revolutions per minute
SNK	Student Newman Keuls
TE	tris-EDTA
TEMED	tetramethylethylenediamine
TGRC	The C.M. Rick Tomato Genetics Resource Center
Tris	trishydroxymethylaminomethane
UV	ultraviolet
v/v	volume per volume
w/v	weight per volume

1 Introduction

1.1 Tomato

The tomato (*Solanum lycopersicum* L.) – originated in middle- and southern America with distributions of its wild relatives from Chile to Venezuela (Warnock, 1991) – is an annual plant and belongs to the solanaceae family. By now it is agreed on *Solanum lycopersicum* var. *cerasiforme* as the most likely ancestor of the tomato cultivars grown today (Costa et al., 2005). Its worldwide distribution started in the 16th century primarily to Europe and around 100 years later the tomato reached Asia (Costa et al., 2005).

In the 19th century the tomato started its triumphal course and today it is the most important and valuable vegetable cultivated all over the world, in the open and under protective cover (Scholberg et al., 2000). In 2006, the worldwide production mounted up to 125,543,475.30 tons (FAOSTAT, © FAO Statistics Division 2007, 04.02.2008) whereof 6,699,000 tons were produced in Asia.

1.2 Heat stress

According to Wahid et al. (2007b) heat stress is defined as the rise in temperature beyond a threshold level for a period of time sufficient to cause irreversible damage to plant growth and development while a transient elevation in temperature, usually 10–15 °C above ambient, is considered heat shock. However, heat stress is a complex function of intensity (temperature in degrees), duration, and rate of increase in temperature. The extent to which it occurs in specific climatic zones depends on the probability and period of high temperatures occurring during the day and/or the night.

Transitory or constantly high temperatures cause various morphological, physiological and biochemical changes in plants affecting plant growth and development and lead to profit cuts (Wahid et al., 2007b). For instance, heat stress influences seed germination negatively (Camejo et al., 2005), has adverse impacts of photosynthesis (Karim et al., 1997; Zhang et al., 2005), respiration (Stone, 2001), water relations (Morales et al., 2003) and membrane stability (Camejo et al., 2005). Moreover, modulations of hormone levels (Maestri et al., 2002), primary and secondary metabolites (Rivero et al., 2001), as well as enhanced expressions of heat shock related proteins (Feder et al., 1999; Schoeffl et al., 1999) and production of reactive oxygen species (Havaux 1998; Sairam et al., 2004) were found as plant reactions to elevated temperatures.

Heat stress affects plants throughout their ontogeny and the responses of several species to temperatures above optimum were investigated in many crops revealing that the sensitivity to heat stress of different plant species is varying tremendously depending on the stage of plant development. In rice, for example, the vegetative growth increased with increasing temperatures (28 to 34 °C, (Baker et al., 1992)) but the grain yield was reduced. The higher biomass accumulation might be advantagous for leafy crop production but production of grains and fruit crops are negatively affected (Wahid et al., 2007b). Reduced plant growth and fruit set in tomato under heat stress are shown in Figure 1.

Figure 1: The heat tolerant variety FMTT260 grown in a greenhouse in Thailand during the dry season. Plant growth and fruit set were reduced, plants showed severe physical disorders, e.g. involute leaves and uneven ripening of the fruits.

1.3 Heat stress in tomato

Although tomatoes are adapted to various climates, their growth and development is rather sensitive to environmental stresses including heat (Foolad, 2005). Under tropical and subtropical climates heat stress is a severe constriction for tomato crop production (Kleinhenz et al., 2006) since it is leading to poor fruit set and consequently low yields.

Various experiments on heat stress in tomato have been conducted already and most of them under short time heat shock treatments. Heat shock means a short time exposure of plants to temperatures significantly above the threshold of plant growth. Heat shock experiments were conducted by Iwahori (1965), Rudich et al. (1977), and Camejo et al. (2005) amongst others.

From the later 1990ies experiments focused more on moderately elevated temperatures (Peet et al., 1997; Peet et al., 1998a; Sato et al., 2000; Sato et al., 2002) with regard to discussions on global warming.

Several traits shown to be affected by high temperatures were reported to be correlated with reduced fruit set in tomato. For instance changes in the carbohydrate supply of different plant organs (Atherton et al., 1986), hormonal imbalances (Kuo et al., 1984), and malfunctions of reproductive organs (Charles et al., 1972; Dane et al., 1991; Peet et al., 1996).

The flower development was shown to be exceptionally sensitive at three different stages. High temperature is most deleterious when flowers are first visible and sensitivity continues for 10–15 days. Reproductive phases most sensitive to high temperature are gametogenesis (8–9 days before anthesis) and fertilization (1–3 days after anthesis, Foolad, 2005).

During reproduction, a short period of heat stress can cause significant increases abortion of floral buds and of opened flowers but great variations in sensitivity within and among plant species were reported (Guilioni et al., 1997; Young et al., 2004).

Impairment of pollen and anther development by elevated temperatures is another important factor contributing to decreased fruit set under high temperatures (Peet et al., 1998; Sato et al., 2006).

Both, male and female gametophytes were shown to be sensitive to high temperature but responses varied with genotypes. In general, ovules were less heat sensitive than pollen (Peet and Willits, 1998).

Nevertheless, most researchers found that poor fruit set is not only caused by a single factor (Rudich et al., 1977; Kuo et al., 1979). Though poor fruit set at high temperature cannot be explained by only one single factor, decreases in pollen germination and/or pollen tube growth are among the factors most commonly reported (Wahid et al., 2007b). In tomato, reproductive processes adversely affected by high temperature included meiosis in both, male and female organs (Kinet et al., 1997), pollen germination and pollen tube growth (Weaver and Timm, 1989), ovule viability (Kinet et al., 1997), stigmatic and style positions (Charles et al., 1972; El-Ahmadi et al., 1979), number of pollen grains retained by the stigma, fertilization and post-fertilization processes, growth of the endosperm, pre-embryo and fertilized embryo (Kinet et al., 1997; Peet et al., 1998b). Also, the most noticeable effect of high temperatures on reproductive processes in tomato is the production of an exserted style, a stigma elongated beyond the anther cone, which may prevent self-pollination (Rick et al., 1969).

Heat tolerance is generally defined as the ability of plants to grow and produce economic yield under high temperatures (Foolad, 2005). The great variation in sensitivity within and among plant species and in responses to heat stress indicates the genetic variability concerning this trait. But in some plant species, for example soybeans and tomatoes, limited genetic variations exist within the cultivated species necessitating identification and use of wild accessions (Foolad, 2005).

Genetic variation in tomato genotypes regarding its heat tolerance was already reported under field conditions (Dane et al., 1991). Therefore, in the current study the response to heat stress of *Solanum lycopersicum,* three wild relatives (*S. pennellii, S. habrochaites,* and *S. pimpinellifolium*), and *Solanum pennellii* introgression lines were investigated in climate chambers and under greenhouse conditions in the tropics. The main focus was laid on traits related to fruit set, especially on pollen characteristics.

Phenotypic data of a segregating population grown under heat stress were used for the combination with data obtained by molecular marker analyses. Both are necessary for linkage mapping and QTL mapping. Since they allow a quick scan of the whole genome for polymorphisms without the need of prior sequence information the AFLP analyses (Vos et al., 1995) were chosen for marker analyses. This method generates large numbers of bands which are highly reproducible.

Since the development of a suitable greenhouse design for tomato production in the tropics was one of the major objectives of the 'Protected Cultivation Project', the program in whose framework this study was accomplished, the influence of different cooling methods introduced in the crop production in the lower latitudes on traits related to fruit set was evaluated in this study.

2 Materials and methods

2.1 Plant material

Table 1: Accessions, scientific names, origins and classification of heat tolerance of the tomato lines and wild species used for the experiments.

Accession	Species	Origin	Heat tolerance
ChiaTai	*Solanum lycopersicum* L.	local variety, Thailand	unknown
CL5915-93D4-1-0-3	*S. lycopersicum* L.	AVRDC	yes
CLN1621L	*S. lycopersicum* L.	AVRDC	yes
CLN2001A	*S. lycopersicum* L.	AVRDC	yes
CLN2418A	*S. lycopersicum* L.	AVRDC	yes
Donna091	*S. lycopersicum* L.	local variety	unknown
FMTT260	*S. lycopersicum* L.	AVRDC	yes
FMTT269	*S. lycopersicum* L.	AVRDC	yes
HT7	*S. lycopersicum* L.	Vietnamese variety	yes
LA2661	*S. lycopersicum* L. cv. Nagcarlang	TGRC	yes
LA2662	*S. lycopersicum* L. cv. Saladette	TGRC	yes
LA3120	*S. lycopersicum* L. cv. Malintka-101	TGRC	yes
LA3320	*S. lycopersicum* L. cv. Hotset	TGRC	yes
LA0716	*S. pennellii* L.	TGRC	unknown
LA1589	*S. pimpinellifolium* L.	TGRC	unknown
LA1777	*S. habrochaites* S. Knapp & D. M. Spooner (form. *L. hirsutum* Dunal)	TGRC	unknown
Pannovy	*S. lycopersicum* L.	Syngenta	no
Sida013	*S. lycopersicum* L.	local variety	unknown
Valentine	*S. lycopersicum* L.	local variety	unknown

AVRDC= The Asian Vegetable Research and Development Center, Shanhua, Taiwan, TGRC= The C.M. Rick Tomato Genetics Resource Center, Davis, California, USA, Syngenta= Syngenta Seeds GmbH, Kleve, Germany)

Table 2: Accession numbers and names of the *Lycopersicum pennellii* introgression lines supplied from the TGRC (The C.M. Rick Tomato Genetics Resource Center) used in the experiments

LA4028 IL1-1	LA4047 IL3-5	LA4062 IL6-3	LA4084 IL9-3
LA4031 IL1-2	LA4048 IL4-1	LA4063 IL6-4	LA4087 IL10-1
LA4032 IL1-3	LA4050 IL4-2	LA4064 IL7-1	LA4089 IL10-2
LA4033 IL1-4	LA4051 IL4-3	LA4065 IL7-2	LA4091 IL10-3
LA4037 IL2-2	LA4053 IL4-4	LA4066 IL7-3	LA4092 IL11-1
LA4038 IL2-3	LA4054 IL5-1	LA4067 IL7-4	LA4093 IL11-2
LA4039 IL2-4	LA4055 IL5-2	LA4069 IL7-5	LA4094 IL11-3
LA4040 IL2-5	LA4056 IL5-3	LA4071 IL8-1	LA4095 IL11-4
LA4041 IL2-6	LA4057 IL5-4	LA4074 IL8-2	LA4097 IL12-1
LA4043 IL3-1	LA4058 IL5-5	LA4076 IL8-3	LA4099 IL12-2
LA4044 IL3-2	LA4059 IL6-1	LA4078 IL9-1	LA4100 IL12-3
LA4046 IL3-4	LA4060 IL6-2	LA4081 IL9-2	LA4102 IL12-4

2.2 Growth conditions

2.2.1 Hannover

Plants were either sown in rock wool (Grodan BV, KD Roermond, The Netherlands) or in tray substrate (Klasmann-Deilmann GmbH, Geeste-Groß Hesepe, Germany) at pH 5.5 (CaCl$_2$, v/v 1:2.5): nitrogen (180 mg N L^{-1}), phosphorus (210 mg P$_2$O$_5$ L^{-1}), potassium (240 mg K$_2$O L^{-1}), and magnesium (120 mg Mg L^{-1}). Soil cultured seedlings were transplanted in black 5 liter pots filled with Potgrond P substrate (Klasmann-Deilmann GmbH, Geeste-Groß Hesepe, Germany), at pH 5.5 (CaCl$_2$, v/v 1:2.5): nitrogen (210 mg N L^{-1}), phosphorus (240 mg P$_2$O$_5$ L^{-1}), potassium (270 mg K$_2$O L^{-1}), and magnesium (120 mg Mg L^{-1}) and nurtured at 24/ 20 °C.

Fertigation was accomplished with flory® 2 mega (Euflor GmbH, München, Germany) with every irrigation at a concentration of 1.5 ‰. Concentrations of the nutrients are listed in the appendix.

The insecticide Plenum® 50 WG (Syngenta Agro GmbH, Maintal, Germany) was applied at a concentration of 0.02 % against whiteflies.

Indeterminate growing plants were pruned once or twice weekly while determinate growing plants remained unpruned. All plants were grown on strings to ensure stability and laid down according to necessity. After the first harvest, senescent leaves were removed regularly up to the first fruit-carrying truss.

2.2.2 Thailand

Seeds were sown in peat moss and kept in an evaporative cooled nursery for two weeks prior to transplanting. After nurturing seedlings were transplanted in white 10 L planting pots filled with a locally purchased substrate (Dinwondeekankasat, Ayutthaya, Thailand). The inorganic portion consisted of 30 % sand, 39 % silt, and 31 % clay, the proportion of organic matter was 28 % and pH was 5.3.

The insecticides Abamectin™ (1.5 ml L^{-1}), Spinosad™ (1.5 ml L^{-1}), and Cypermethrin™ (2 ml L^{-1}) were alternately sprayed on a weekly basis and the fungicide Mancozeb™ (4 ml L^{-1}) was applied once before the start of flourishing.

Fertigation was done automatically by single dripper irrigation. The nutrient composition of the fertigation solution is listed in the appendix.

Indeterminate growing plants were pruned twice weekly while no pruning took place in determinate growing varieties. In order to ensure stability, all plants were grown using a high wire growing system as described in detail by (Kleinhenz et al., 2006). Plants were laid down according to necessity. After the first harvest, senescent leaves were removed regularly up to the first fruit-carrying truss.

2.3 Greenhouse and climate chamber facilities

2.3.1 Climate chambers (CCs)

Experiments conducted in climate chambers (CC) of the Department of Horticulture in Hannover, Germany, were run under controlled conditions of temperature, air humidity and light. The size of the CC was 320 × 250 cm (L × W).

The temperature regime was 34/ 30 °C (heat stress) and 24/ 20 °C (optimum temperatures) day/ night lasting for 14 and 10 hours (h), respectively. In the morning and the evening sunrise and sunset were imitated by lighten up and dim down the lamps slowly for one hour, respectively. The regular daytime light intensity was 700 µE m^{-2} s^{-1} and targeted relative humidity (RH) was 60 %.

2.3.2 Greenhouse cabinets (GCs)

The size of the GCs at the Department of Horticulture in Hannover, Germany, was $3 \times 3 \times 2.5$ m (L × W × H) or $8 \times 12 \times 4$ m (L × W × H). The temperature regime was 32/ 28 °C (heat stress) and 24/ 20 °C (optimum temperature) during day/ night, respectively. Artificial light (sodium discharge lamps) illuminated the cabinets in early morning hours and late afternoon resulting in 16 h lasting days.

2.3.3 Greenhouses (GHs)

The GHs at the experimental facilities of the 'Protected cultivation project' were situated on the campus of the Asian Institute of Technology (AIT), Klong Luang, Pathum Thani, Central Thailand (14° 04' N, 100° 37' E, altitude 2.3 m). GHs of two different sizes were used for the experiments. The dimensions of the big GHs were $10 \times 20 \times 6.4$ m (L × W × H). The roof and the lower parts of the sidewalls were mounted with an 200 µm UV-absorbing polyethylene (PE) film (Wepelen™, anti-dust, anti-fog FVG, Dernbach, Germany) up to a height of 50 cm while the remaining part of the sidewalls as well as the roof vent (height: 0.8 m) were clad with 50mesh insect-proof net (BioNet, Klayman Meteor Ltd, Petach-Tikva, Israel).

The size of the small GHs was $6.0 \times 3.0 \times 3.2$ m, (L × W × H). The sidewalls of these GHs were clad with 40mesh net (Econet M, Ab Ludvig Svensson, Kinna, Sweden) and the aforementioned Wepelen™ PE-film was used as roof cover.

All GH floors were covered with a bi-colored (black/ white) plastic mulch (Silo plus™, FVG, Dernbach, Germany).

For some experiments the roof plastics were coated with a near infrared (NIR) reflecting pigment paint (Reduheat, Mardenkro, The Netherlands, mixing ratio 1:2.5 pigment to water).

One GH was entirely clad with PE film and equipped with an evaporative ('fan and pad') cooling system (FAP). Beside the FAP-GH all big greenhouses were operated with combination of natural and mechanical ventilation the latter provided by two exhaust fans (∅ 1 m, capacity: 550 m^3 min^{-1}) which were switched on automatically whenever inside air temperatures exceeded 32 °C. The small GHs were ventilated naturally.

2.4 Evaluation of phenotypic traits

Inflorescences, flowers and fruits were counted from bottom to top and from the stem to the tip of the inflorescence and fruits scored set when they reached a diameter of at least 0.5 cm.

Only flowers developed and opened completely were recorded.

The height of the indeterminate growing plants was measured from the surface of the substrate in the pot to the shoot tip. Of the determinate growing plants, always the longest shoot combination of the first stem and the diverging stems were measured.

The pollen of single flowers were collected into reaction tubes (Sarstedt, Nümbrecht, Germany) by shaking their pedicel with an electrical toothbrush (Oral-B, Braun GmbH, Kronberg, Germany) and counted via transmitted-light microscopy (Axiovision40, Zeiss, Göttingen, Germany) with tenfold magnification by using of a Fuchs-Rosenthal counting chamber (Carl Roth GmbH + Co. KG, Karlsruhe, Germany).

For testing the pollen viability the staining procedure according to Heslop-Harrison et al. (1984) using FDA was applied.

2.5 Chemicals

2.5.1 Pollen staining solutions

FDA staining

For the stock solution, 2 mg FDA (Fluorescein diacetate [3',6'-Bis(acetyloxy)-spiro isobenxofuran-1(3H),9'-9H xanthen-3-one 3,6-Diacetoxyfluoran], Sigma-Aldrich Chemie GmbH, München, Germany) were solved in 1 ml acetone. The stock solution was mixed with a 10 % [w/ v] sucrose solution.

Aniline blue staining

The fixer consisted of ethanol (96 %) and lactic acid (90 %) at a ratio of 2:1.

The aniline blue staining solution was prepared in DI (deionized water) with 1.36 mM aniline blue and 36.16 mM $K_3PO_4 \times H_2O$ and exposed to natural light until the color changed from blue to yellow but at least for 24 hours.

MTT staining

The MTT (3-(4,5-dimethylthiazol-2-yl)-2,5-diphenyl tetrazolium bromide, Sigma-Aldrich Chemie GmbH, München, Germany) staining solution consisted of 1 % (w/ v) MTT in a 5 % (w/ v) sucrose solution according to Rodriguez-Riano et al. (2000).

2.5.2 Pollen growth media

Table 3: Components and their concentrations in the liquid pollen growth media used in the experiments. The names of the media (except 'sucrose + boric acid') are according to their publishing authors.

Medium	Brewbaker and Kwack	Heslop-Harrison	Poulton	sucrose + boric acid
Components	Concentration [mM]			
sucrose	0.292	0.351	0.409	0.292
H_3BO_3	1.617	1.000	1.617	1.617
$Ca(NO_3)_2 \times 4\, H_2O$	1.828	1.000	1.828	
$MgSO_4 \times 7\, H_2O$	1.662		1.662	
KNO_3	0.989		0.989	

2.6 Histological techniques

The fixer for histological investigations of flowers – AFE – consisted of Ethanol (96 %), formaldehyde, and acetic acid in the ratio of 18:1:1.

The infiltration for the embedding took place in the infiltration solutions for 16 hours according to a modified Kulzer protocol (Technovit 7100, 01/ 04, Table 4 to Table 11).

Table 4: Composition of the pre-infiltration solution used before embedding gynoecia and androecia in synthetic resin. (All ingredients except ethanol were supplied by Heraeus Kulzer GmbH, Werheim/Ts., Germany).

Pre-infiltration solution
Technovit7100 Stock solution
96 % undenaturated ethanol
ratio 1:1

Materials and Methods

Table 5: Composition of the infiltration solution used before embedding the gynoecia and androecia in synthetic resin. (All ingredients were supplied by Heraeus Kulzer GmbH, Werheim/Ts., Germany).

Infiltration solution
100 ml Technovit7100 Stock solution
1 g Härter1

Table 6: Composition of the embedding synthetic solution used for embedding the gynoecia and androecia. (All ingredients were supplied by Heraeus Kulzer GmbH, Werheim/Ts., Germany).

Embedding synthetic
15 ml infiltration solution
1 ml Härter2

Table 7: Composition of the synthetic resin solution used for attaching the grips on the blocks. (All ingredients were supplied by Heraeus Kulzer GmbH, Werheim/Ts., Germany).

Synthetic resin
Technovit3040 powder
Technovit3040 solution
ratio 3:1

Grips attached with the blocks were clamped in the rotary microtome (Reichert-Jung, now: Leica Biosystems Nussloch GmbH, Nussloch, Germany) and the cuttings were done with tempered blades (HistoknifeH, Heraeus Kulzer GmbH, Werheim/Ts., Germany).

Cuttings of the anthers were stained with methylene blue or hematoxilin and cuttings of ovaries with methylene blue, hematoxylin according to Delafield (Gerlach 1984), or with Periodic acid-Schiff stain (PAS, Feder et al., 1968). For staining substances and the preparation of the staining solutions see Table 8.

Flowers at early stages were covered with Entellan® (Merck KGaA, Darmstadt, Germany).

Table 8: Components of staining solutions used for the staining procedures of the microtome cuttings of androecia and gynoecia, their preparation and concentrations. Substances labeled with * were supplied by Carl Roth GmbH + Co. KG, Karlsruhe, Germany, and labeled with + by Sigma-Aldrich Chemie GmbH, München, Germany.

Components	Preparation	concentration [mM]
methylene blue staining solution*	solved in DI	15.63
hematoxylin solution according to Delafield+		pure
dimedone+	solved in DI, stirred for 5 hrs, filtrated	35.67
periodic acid+	solved in DI	43.87
potassium metabisulfite+	solved in DI	19.12
pararosalinin (in 1N HCl)+	mixed with potassium metabisulfite, placed dark for 24 hrs, mixed with 0.5 g active carbon, agitated for 2 min, and filtrated	95.95

2.7 Microscopical techniques

Two different microscopes were used: Axiovision40 and Axiovision20 (both Zeiss GmbH, Göttingen, Germany) both with UV light illumination accomplished by high-pressure mercury lamps HBO50W (Osram, München, Germany).

Pictures were taken by a preinstalled digital camera (PowerShot A70, Canon Deutschland GmbH, Krefeld, Germany).

FDA stained pollen were evaluated under the microscope using the filter set 09 (Zeiss GmbH, Göttingen, Germany) with excitation at 450- 490 nm and emission at 515 nm.

Aniline blue stained samples were evaluated using the filter set 02 (Zeiss, Göttingen, Germany) with excitation at 365 nm and emission at 420 nm.

2.8 Flow cytometry

For the evaluation of pollen viability a flow cytometer EPICS XL-MCL (Beckman Coulter GmbH, Krefeld, Germany) was used. Every sample was investigated for 180 seconds. Results measured with the flow cytometer were displayed using the computer SYSTEM II™ software (Beckman Coulter, Inc., Florida, USA).

2.9 DNA extraction

Dried leaf material was ground in 2 ml reaction tubes (Eppendorf AG, Hamburg, Germany). DNA was extracted according to a protocol of Engel (personal communication, 2005, for details see appendix). 10 mg of ground material were mixed with 400 µl extraction buffer and mixed thoroughly. The samples were incubated in a water bath at 65 °C for 30 min. 500 µl chloroform were added, mixed, and spun down for 5 min at 13,000 rpm in a micro centrifuge (5415D, Eppendorf, Hamburg, Germany). The supernatant was transferred to a new tube and mixed with 600 µl CTAB buffer. After swaying the solution, it was incubated for 15 min at room temperature and swayed again subsequently. The mixture was spun down at 13,000 rpm for 15 min. After centrifugation and pellet agglutination the supernatant was discarded and the pellet was dried. The pellets were dissolved in 600 µl TE high salt. 800 µl ice-cold ethanol were added and mixed. After centrifugation at 13,000 rpm for 5 min, the pellet was dried. For storage, the DNA was dissolved in 200 µl TE 01. The DNA solution was stored at -4 °C. For compositions and concentrations of the used buffers, see appendix.

2.10 DNA quantification

10 µl of DNA solution were mixed with 1 µl Orange G loading buffer (AppliChem GmbH, Darmstadt, Germany) and applied on a 1 % agarose gel (PEQLAB Biotechnologie GmbH, Erlangen, Germany) with 0.01 % ethidium bromide (Carl Roth GmbH + Co. KG, Karlsruhe, Germany).

λ-DNA was applied on the gel in different concentrations ranging from 10 ng to 100 ng. The λ-DNA at different concentrations served as a standard to generate the calibration curve. The DNA quantity of the samples was compared to the curve. The electrophoresis was run at 80 V for approx. 30 min. A Gel iX Imager with a 312 nm UV light emitting table was used to visualize the ethidium bromide labeled DNA. A picture of the agarose gel was taken with the preinstalled camera (both Intas Science Imaging Instruments GmbH, Göttingen, Germany). The DNA amounts of the samples were determined using the Gel-Pro Analyzer (version 4.5.00.0, Media Cybernetics, Inc., Bethesda, USA).

2.11 AFLP Analyses

AFLP analyses were done according the protocols of von Malek et al. (2000), further on referred to as protocol 1) and Truong (2007), further on referred to as protocol 2) with small modifications. The latter one is appended.

2.11.1 Restriction

DNA was restricted using different enzyme combinations. Two restriction enzymes were used simultaneously: *Hin*dIII or its homolog *Tru*I, rare cutting restriction enzymes with six bp recognition sites, and *Mse*I, a frequent cutting restriction enzyme with a four bp recognition site. The recognition sites are A|AGCTT for *Hin*dIII/ *Tru*I and T|TAA for *Mse*I.

The second enzyme combination consisted of *Eco*RI (recognition site G|AATTC) and *Mse*I. The enzymes and buffers were provided by New England Biolabs, Ipswich, Great Britain.

The restrictions took place at 37 °C and their correctness was controlled on 1 % agarose gels. Completely digested samples showed a smear of 100- 1000 bp length.

A mixture containing two restriction enzymes (*Eco*RI and *Mse*I) was used to digest a total amount of 200 ng DNA (Table 9). The restriction reaction was incubated overnight at 37 °C in an incubator or for four hours in a heating block. 10 µl of the digested product were load on 1 % agarose gel with 10 µl of a 100 bp ladder. Entirely digested samples showed a smear of 100- 1000 bp length. The digestion products of *Eco*RI/ *Mse*I were incubated at 70 °C for 15 min to stop the enzyme reaction

Table 9: Components, concentrations and volumes used for the restriction cocktail according to protocol 2.

Component	Concentration	Volume [µl/ tube]
DNA	10 ng/ µl	20.0
H$_2$O		9.8
10 x buffer 2 (BioLabs)		3.0
EcoRI (BioLabs)	20 U/ µl	0.8
MseI (BioLabs)	10 U/ µl	1.2
10 x BSA (BioLabs)		0.2
Total		35.0

Ligation

10 µl of a adaptor containing cocktail (Table 10) were added to 20 µl of the restricted DNA and incubated overnight at 16 °C or over weekend at 12 °C or in the thermocycler for 4 hrs at 37 °C. The ligase and its corresponding buffer were provided by Fermentas GmbH, St. Leon-Rot, Germany.

Table 10: Components, concentrations and volumes used for the ligation cocktail according to protocol 2.

Component	Concentration	Volume [µl/ tube]
H$_2$O		6.6
10 x ligation buffer		1.0
EcoRI adaptor (MWG)	5 µM	1.0
MseI adaptor (MWG)	50 µM	1.0
T4 DNA ligase	400 U/ µl	0.4
Total		10.0

The ligation product was spun down and diluted 1/10 in DI. Samples were stored at -20 °C.

Pre-selective amplification

5 µl of the digested and ligated product were mixed with a cocktail (Table 11) containing two pre-selective primers without selective nucleotide extension at the ends. A PCR was conducted using the thermocycler TGradient (Whatman Biometra GmbH, Göttingen, Germany). The PCR conditions are listed in Table 12.

The pre-amplification products were loaded on a 1 % agarose gel to test the reaction. In case of a successful reaction a smear of 50 to 500 bp was clearly visible.

Table 11: Components, concentrations and volumes used for the pre-amplification cocktail according to protocol 2.

Component	Concentration	Volume [µl/ tube]
H_2O		9.6
10x Williams buffer		2.0
*Eco*RI primer (Eco+0, MWG)	10 µM	0.6
*Mse*I primer (Mse+0, MWG)	10 µM	0.6
dNTPs	2 mM	2.0
Taq polymerase (Firepol)	5 U/ µl	0.2
aliquot of digested and ligated DNA		5.0
Total		20.0

Table 12: PCR conditions for the pre-amplification according protocol 2.

Step	Time [sec]	Temperature [°C]
Denaturation	30	94
Annealing	30	56
Extension	60	72
28 cycles		

Pre-amplified products were diluted 1:20 and were stored at -20 °C (diluted reaction products could be stored at 4 °C for daily use).

Selective amplification

The pre-amplified DNA products were mixed with the amplification cocktail (Table 13) and another PCR conducted under conditions as listed in Table 14.

Table 13: Components, concentrations and volumes used for the amplification cocktail according to protocol 2.

Component	Concentration	Volume [µl/ tube]
H_2O		4.78
10 x Williams buffer		1.00
Eco-IRD700 primer * (MWG)	0.5 µM	0.75
*Mse*I primer (MWG)	8.52 µM	0.43
dNTPs	2 mM	1.00
Taq polymerase (Firepol)	5 U/ µl	0.05
aliquot of pre-amplified DNA diluted 1:30		2.50
Total		10.0

The cocktail contained two selective primers with three nucleotide extensions each. The final product was diluted 1:4 with the AFLP loading buffer (see appendix).

Table 14: PCR conditions for the selective amplification according to protocol 2.

Step	Time [sec]	Temperature [°C]
Denaturation	30	94
Annealing	30	65
Extension	60	72
1 cycle		
Denaturation	30	94
Annealing	30	65 (-0.7 °C in every cycle)
Extension	60	72
12 cycles		
Denaturation	30	94
Annealing	30	56
Extension	60	72
28 cycles		

Polyacrylamide gel electrophoresis (PAGE)

For the gel electrophoreses 25 cm long and 0,25 mm thick 7 to 9.85 % denaturing polyacrylamide gels were prepared. 8.4 g Urea (Diaminomethanal, USB Corporation, Cleveland, Ohio, USA) were mixed with 10.24 ml DI and 4 ml acrylamide/ bis- solution (Rotiphorese® NF-acrylamide/bis-Lösung 40 % (19:1), Carl Roth GmbH + Co. KG, Karlsruhe, Germany) was added. 2 ml Longrun TBE (composition see appendix) were added and stirred thoroughly until the Urea was dissolved entirely. 140 µl ammonium persulfate (APS, Sigma-Aldrich Chemie GmbH, München, Germany) and 20 µl TEMED (Tetramethylethylenediamine, AppliChem GmbH, Darmstadt, Germany) were added under continued stirring. The solution was poured quickly between two glass plates avoiding the formation of air bubbles. After polymerization 0.8 µl of the final amplification products samples mixed with the loading buffer 1:3 were loaded on the gel. The gels were run on a DNA Analyzer (Gene readir4200, LI-COR Biosciences GmbH, Bad Homburg, Germany) using the software Liquor eSeq (version 2.0.38, LI-COR Corporate Offices, Nebraska, USA).

2.12 Data analyses

In experiments with more than two plants per genotype, analyses of variances (ANOVA) were conducted for all traits investigated per plant ensuring no misinterpretation of the results evoked by inhomogeneous plant material. Where no differences between plants of one genotype were found the data were pooled.

For all experiments multi-factorial ANOVAs were calculated depending on the number of factors. SAS's Version 9.1 (SAS'S, 2005, SAS'S Institute Inc., Cary N.C., USA) was used for all statistical analyses. Except where explicitly indicated otherwise, a significance level of $\alpha = 0.05$ was used for all comparisons of means.

For all experiments in which inflorescences, flowers, and fruits were counted these factors were assumed independent from each other. The data were considered for statistical analyses when $n \geq 3$.

Fruit set was calculated as the quotient of the flower number and fruit number. The percentage of fertilized fruits was calculated as the proportion of seeded fruits and total fruit number.

In experiments with genotypes of different growth habits (determinate/ indeterminate) the weekly increment of the plant height was used for the analyses since the total plant height always differed significantly between these two groups.

Means of data sets with inhomogeneous or homogeneous sample sizes were compared using the Student Newman Keuls (SNK) test or the Tukey test, respectively. Data sets with high sample sizes were additionally adjusted according to Bonferroni.

In those cases where the pollen amount was not normal distributed the data were logarithmized before analyzed regarding to Sachs (1993). After analysis, the data were retransformed back to their original scale.

To analyze the percentage data of pollen viability the data set was transformed using an arcos sinus transformation. Pollen viability data were considered for analyses when at least 100 pollen grains per flower and sample could be evaluated.

3 Results

3.1 Highest applicable temperature for heat stress experiments

The experiment was performed to detect the highest temperature applicable for conducting investigations of heat stress in tomato. The temperature should be high enough to induce heat stress symptoms within the plants but must not exceed a limit beyond which retarded plant growth or inhibition of all physiological processes are caused. Seeds of FMTT260 and Pannovy were sown in rock wool (Grodan BV, KD Roermond, The Netherlands) in calendar week (cw) 50 in 2004. After three weeks (cw1/2005) of nurture at optimum temperatures, thirty-two plants were transplanted into two climate chambers (CCs) resulting in a plant density of approx. four plants m^{-2}. Eight plants of both hybrids were placed randomly in metal tubs (87 × 120 cm) on four tables (Figure 2). The metal tubs had an incline to support the runoff of the leachate. This solution was collected in a central tank, pumped back to the metal tubs, and dispersed to the single plants by dripping irrigation.

After four weeks of heat treatment (34/ 30 °C) the plants had developed several physical disorders. While leaves of plants grown under optimal temperature (24/ 20 °C) were faintly involute, leaves developed under heat stress were characterized by extremely erected and severely involute leaf blades (Figure 3, Figure 4a). Flowers developed under heat stress of both varieties showed severe abnormalities, e.g. loss of the reproductive organs and stigma exsertion (Figure 4b and c). Under heat stress, several flowers dropped before they opened or they withered after opening without reaching the anthesis stage. Even under optimal temperatures buds and flowers of the hybrid Pannovy were characterized by elongated stigmata protruding over the anthers (Figure 5) while no such symptoms were observed in FMTT260.

Under optimum temperatures all four experimental plants of Pannovy and one plant of FMTT260 started flowering in cw4/2005 at the age of six weeks. The remaining plants of FMTT260 started flourishing in cw5/2005. The first fruit set was observed in cw5/2005 in all plants of Pannovy and in one of FMTT260.

Results

Figure 2: Construction of the climate chambers used and distribution of the tomato plants of the varieties FMTT260 and Pannovy on the transplanting day. Four plants per mat and eight plants per table (completely randomized) were grown on rock wool, fertigation was done automatically.

Under heat stress, three plants of Pannovy and one of FMTT260 started flowering in cw4/2005 and the remaining four plants one week later. The first fruit set was observed in cw5/2005 in two plants of Pannovy, followed by three plants of FMTT260 in cw6/2005. Within the course of the experiment plants of Pannovy developed around seven and five, those of FMTT260 five and four inflorescences at optimum temperatures and under heat stress, respectively.

After 4 weeks, no vital inflorescences were developed in the heat treatment anymore and the experiment had to be ceased.

Results

Figure 3: Tomato plants (Pannovy and FMTT260) grown under heat stress (34/ 30 °C, night/ day, a) or under optimum temperature (24/ 20 °C, b) conditions for 4 weeks in climate chambers at Hannover University.

Figure 4: Heat stress symptoms of tomato plants grown at 34/ 30 °C (day/ night) (in climate chambers at Hannover University): a) Involute and stiff leaves, b) flowers without any reproductive organ and c) stigma exsertion.

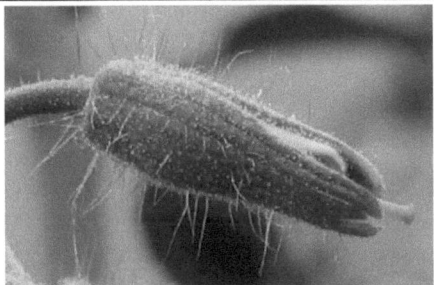

Figure 5: Flower bud of variety Pannovy developed under optimum temperatures (24/ 20 °C) in a climate chamber at Hannover University. The stigma protrudes the petals, sepals and anthers.

The influence of temperature and variety on number of inflorescences, sum of flowers, and fruits (Table 15) counted from cw5/2005 to cw8/2005 from four plants per genotype were highly significant.

The numbers of inflorescences, flowers and fruits were reduced under heat stress and Pannovy always performed better compared to FMTT260.

The sum of inflorescences and flowers were more severely decreased in Pannovy compared to FMTT260.

No significant effect of temperature was observed regarding fruit set, but, however, Pannovy had a higher percentage of set fruits than FMTT260 regardless the temperature treatment. The fruit set was calculated as the quotient of the number of fruits developed from the flowers per inflorescence.

Table 15: Average number of inflorescences (Infav), flowers (Flav) and fruits (Frav) and percentage of set fruits (Frset) of heat tolerant (FMTT260, 'FMTT') and heat sensitive (Pannovy, 'Pa') tomato hybrids under optimum temperatures (24/ 20 °C, 'optimum') heat stress (34/ 30 °C, 'stress'). Different letters within columns indicate significant differences. The two factors did not interact.

Treatment	Infav	Flav	Frav	Frset [%]
Pa/ optimum	6.75 a	55.75 a	39.50 a	71.40 a
FMTT/ optimum	5.00 b	38.50 b	24.00 b	62.60 b
Pa/ stress	4.50 c	34.25 c	28.50 c	84.70 a
FMTT/ stress	3.75 d	29.00 d	16.00 d	56.30 b

Results

The pollen amount released from the anthers (herein further referred to as pollen amount) and pollen viability were analyzed once a week from four plants of each variety.

To evaluate their viability pollen were stained with the FDA solution consisting of the stock solution dispersed in 10 % sucrose solution at a ratio of 1:25. At this concentration a permanent turbidity of the final staining solution occurred. 300 µl of the staining solution were added to the pollen and mixed thoroughly. 20 µl of the pollen suspension were placed on a microscope slide after five minutes incubation time at room temperature. Pictures of at least 100 pollen grains were taken by the preinstalled digital camera to avoid fading of the stain by the UV light while classifying. Viable pollen fluoresced bright under UV-light while non-viable pollen did not fluoresced and therefore only their silhouettes were visible. Some pollen showed a green color but did not glow brightly (Figure 6). The latter were rated in a third group (degraded pollen). The pictures were analyzed at the computer using an image editing software (Axiovision AC, V 4.2.0.0, Carl Zeiss GmbH, Göttingen, Germany).

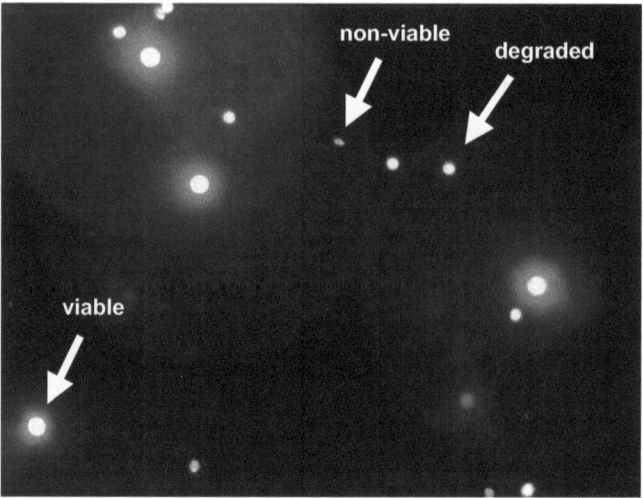

Figure 6: Pollen stained with FDA and excited with UV light glow with different intensity depending on their viability. Viable pollen fluoresce brightly, degraded pollen less bright and non-viable pollen do not glow and often appear misshaped. The picture was taken under the microscope with 100fold magnification and the filter set 09 (Zeiss GmbH, Göttingen, Germany).

To evaluate the pollen amount flowers were shaken for exactly 5 seconds with an electric toothbrush. The samples were dispersed with 300 µl staining solution and thoroughly mixed just before counting using a Fuchs-Rosenthal counting chamber.

Results

The pollen amount was significantly reduced in the heat treatment (Figure 7) while the variety did not influence the pollen release (Figure 7). Therefore, and since no significant interaction between the factors were found, the data were pooled and the means across varieties were used for statistical analyses. Since the number of pollen released under heat stress was too low, no comparison of pollen viability between the temperature treatments was done.

Under optimum temperatures, no differences between the varieties regarding their pollen viability were found (Figure 8).

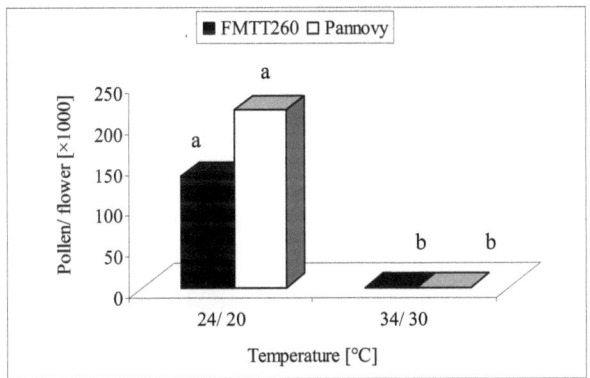

Figure 7: Pollen amounts of the tomato hybrids FMTT260 (heat tolerant) and Pannovy (heat sensitive) under optimal temperature (24/ 20 °C, day/ night) or heat stress conditions (34/ 30 °C). Means with different letters are significantly different (SNK test, α< 0.05, optimum: n= 197; heat stress: n= 62).

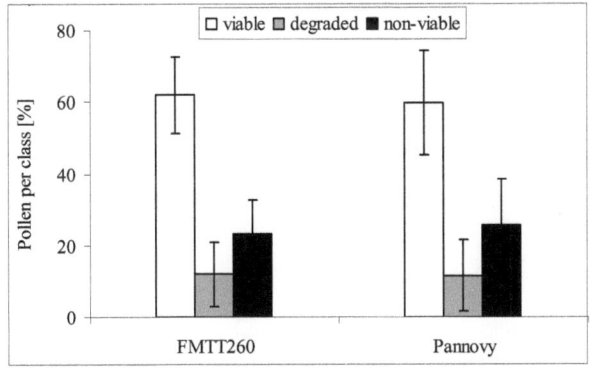

Figure 8: Percentages of viable, degraded and non-viable pollen of the two tomato hybrids FMTT260 (heat tolerant) and Pannovy (heat sensitive) grown in climate chambers under optimum temperatures (24/ 20°C, day/night). Error bars indicate the standard deviation.

Results

3.2 Verification of the integrity of the gynoecium

To check the integrity of the gynoecium four plants of each variety grown under heat stress were pollinated with pollen of plants of the same variety produced under optimum temperatures. Gynoecia unaffected by heat had to be able to develop into fruits after pollination with pollen developed under optimum temperatures supposed to be intact.

Pollen developed under 24/ 20 °C were harvested immediately after anthesis and transferred with a soft brush to stigmata of flowers grown under 34/ 30 °C just after anthesis. Four other plants grown under heat stress were shaken regularly to assure self pollination and comparability. The fruit set was checked once a week and the seed set once by cutting the fruits at the end of the experiment.

In both genotypes, the percentage of fertilized fruits could be increased after hand pollination with pollen produced under optimum temperatures. The percentages (% of total amount of evaluated fruits) of fertilized fruits without additional pollinating (naturally fertilized fruits) were 11 % in FMTT260 and 0 % in Pannovy. After hand pollination, the percentage of fertilized fruits was increased by up to 45 % in FMTT260 and by 5 % in Pannovy (

Figure 9). Though differences between the treatments were clearly visible they were statistically not significant.

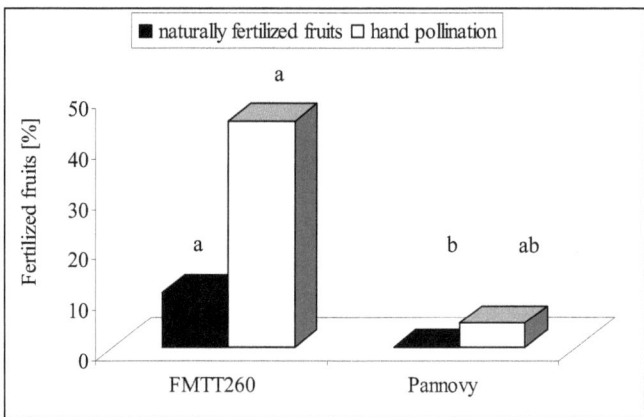

Figure 9: Percentage of fertilized fruits per plant of the two tomato varieties FMTT260 (heat tolerant) and Pannovy (heat sensitive) grown in climate chambers under heat stress (34/ 30°C, day/ night). Flowers were pollinated either by natural fertilization with pollen produced under heat stress or by hand-pollination with pollen produced under optimum temperatures (24/ 20 °C). Means with different letters are significantly different between cultivars and pollination treatments (Tukey test, α < 0.05, n= 14).

Results

3.3 Histological investigations of androecia and gynoecia grown under heat stress

This experiment was conducted within the framework of the Diploma thesis by (Urban 2006) to define the point of time when the pollen development got interrupted or when the pollen lose their functional capability.

Additionally, the gynoecia of flowers grown under two different temperature treatments (heat stress/ optimum temperatures) were investigated to verify the integrity of the gynoecia grown under heat stress.

The varieties Pannovy (heat sensitive) and FMTT260 (heat tolerant) and line CL5915 (heat tolerant) were used for this experiment. Four plants each of every genotype were grown in two GCs under optimum temperatures and heat stress, respectively. In total five sets of plants were sown with a time offset of two weeks and shifted into climate chambers around 10 days before the start of flowering resulting in plant densities between 1.3 and 2.7 plants m^{-2}.

Buds and flowers in all developmental stages were collected for the histological investigations. The sepals and petals were removed and the remains of the flowers fixed with AFE for a minimum of 24 hours. Small buds were fixed completely to avoid damaging by preparation.

For embedding the samples had to be dehydrated and de-aerated. They were placed in alcohol of afferent concentrations (70 %, 90%, 100% denatured alcohol, and 100 % pure alcohol) for two hours each. To facilitate the de-aeration the samples were put under vacuum until no bubbles ascended from the tissue anymore. Successive, the samples were stored in pre-infiltration solution (Table 4) for two hours. Infiltration took place in the infiltration solution (Table 4) for 16 hours.

The embedding was done in Glycidyl methacrylate (GMA) in a Teflon-coated form and polymerization happened at 40 °C for 12 hours. For the embedding in GMA the liquid methacrylate (embedding synthetic, Table 4) was filled into the gaps of the Teflon form until they were filled completely. The samples were placed upright or across in the middle of the cavity. Polymerization happened at 40 °C in a drying cabinet for 12 hours. The blocks could be stored in closed boxes in the fridge for several weeks.

Grips were attached with a rapidly desiccating resin (Table 4) and cuttings were performed with tempered blades in the rotary microtome. Every third cutting each of 7 µm thickness was used for evaluation. While cutting, the thin layers got compressed and had to be relaxed in a warm water bath. After recreation, they were placed on microscope slides and dried at room temperature.

Cuttings of the anthers were stained with methylene blue or with hematoxylin. Ovaries were stained with methylene blue, hematoxylin according to Delafield (Gerlach, 1984), or with Periodic acid-Schiff stain (PAS, Feder et al., 1968).

Results

The staining with methylene blue was done for 20 sec and samples rinsed with DI afterwards.

For the hematoxilin staining the samples were incubated in the staining solution for two to three minutes and rinsed with DI. To obtain a good contrast the pH of the sorrel samples had to be increased by rinsing with tap water. The color changed to violet-blue.

For PAS, the samples were laid in dimedone solution for 16 to 24 hours. They were rinsed thoroughly in DI for six times, five minutes each. The samples were transferred into periodic acid for ten minutes and rinsed ten times with DI afterwards, three minutes each. Samples were stained in Schiff-reagent until they changed their color to red followed by five stages of a rinsing program by 0.5 % sodium disulfide. In a final step the samples were rinsed by DI for five times. Every wash step lasted three minutes. An Axioskop20 (Zeiss, Göttingen, Germany) microscope was used. Pictures were taken with the preinstalled camera.

The characteristics of the pollen in anthers shortly before or at the time of anthesis differed between the temperature treatments. The pollen developed under optimum temperature were round and could be stained evenly independent of the stain used (Figure 10).

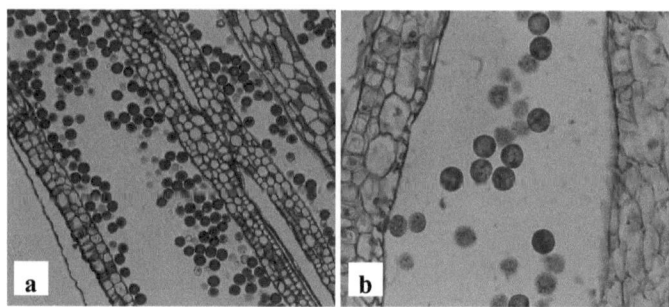

Figure 10: Sections of anthers developed under optimum temperatures stained either with methylene blue (a) or hematoxylin according to Delafield (Gerlach, 1984) (b). Both figures show longitudinal sections in 50fold (a) and 100fold (b) magnification by light microscopy.

The pollen of Pannovy and FMTT260 developed under heat stress looked different: some were round and stained evenly but most looked collapsed and rejected the stain while most of the pollen of the heat tolerant line CL5915 looked round in both temperature treatments.

Although the number of pollen was not systematically quantified, it appeared that Pannovy and FMTT260 produced less pollen under heat stress compared to optimum temperatures.

All anther specific tissues looked normal and the opening process of the stomia was performed in all genotypes and under both temperature treatments (Figure 11). Only the aperture of the stomia was reduced in Pannovy and FMTT260 under heat stress.

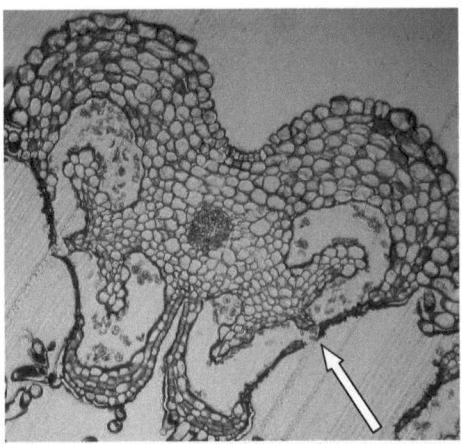

Figure 11: Cross section of an anther of Pannovy developed under heat stress (32/ 28 °C). The tissue was stained with methylene blue and the picture taken by light microscopy with 50fold magnification. The arrow depicts the opened stomium.

To detect the point in time when pollen were damaged during their development, buds and flowers of different developmental stages were investigated. In both temperature treatments, buds of eight different developmental stadia were found: the prothallium, sporogenic cells, meiotic stadium, tetrad stadium, single microspores, pollen and closed septa, pollen and opened septa, and pollen and opened stomia. The later stages of the pollen cell wall development and the mitosis could not be found with these methods. In the early stages, no differences could be found. In the prothallium stage, the septa of the anthers were developed completely. The sporogenic cells of the genotypes did not differ and the anther wall tissues existed. Meiosis was found in all treatments and it proceeded normally. Tetrads were built and their tissues were developed normally. From the tetrads, free microspores were released and they did not differ between the treatments. The tapetum was developed completely. The single anther tissues were developed without disruption and at the right point of time degraded if necessary. Visible differences were only observed in completely developed pollen.

To evaluate the integrity of the gynoecia further on, several staining methods were used to assure the visualization of possibly existent differences in the tissue development between ovaries grown under optimum and elevated temperatures.

No differences in the evaluated ovaries of the different genotypes could be found in the two temperature treatments with different staining methods. The tissues at different developmental stages of the ovaries did not show any variation between the two temperature treatments or the varieties used (Figure 12).

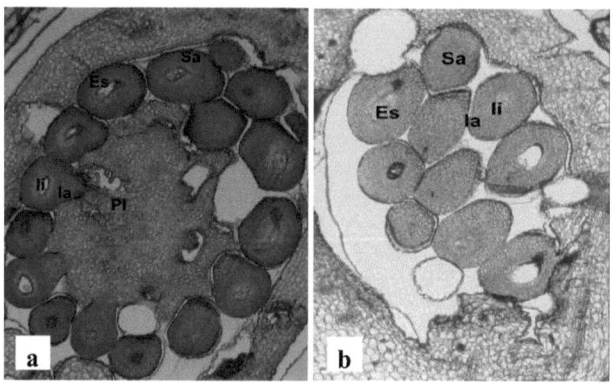

Figure 12: Cross cuttings of ovaries of variety FMTT 260 developed under optimum temperatures (24/ 18 °C, a) and heat stress (32/ 28 °C, b). The cuttings were stained with methylene blue and pictures were taken by light microscopy with 100fold magnification. Es= embryo sack, Ia= outer integument, Ii= inner integument, Sa= ovule, Pl= placenta

3.4 Response of different tomato species and genotypes to heat stress in climate chambers

The experiment was conducted to investigate the heat tolerance within tomato lines known or suspected to be heat tolerant and three wild species. This experiment was undertaken in a CC under heat stress (32/ 28 °C). The seeds were sown in cw4/2005 and two plants of each genotype were transferred to the climate chamber in cw6/2005. Plants were irrigated according to physiological plant stage and necessity. Chemical treatment for plant protection was carried out twice during the experiment.

The number of flowers and fruits were evaluated twice a week beginning in cw12/2005.

The seed set and the production of the colloidal tissue was examined once after the harvest in cw15/2005. In this experiment, the shaking times of the pedicels were raised to ten seconds to qualify the pollen release. The pollen collection, counting, and staining were conducted once a week starting from cw12/2005 to cw15/2005.

Results

Under heat stress conditions, the wild species LA0716 (*S. pennellii* L.), LA1777 (*S. habrochaites* S. Knapp& D. M. Spooner), and LA1589 (*S. pimpinellifolium* L.) did not form any flowers.

The plants of LA2661 developed flowers buds but none of them opened and they shriveled before flowers opened. Some of the flowers of the lines CL5915, LA2662, and FMTT260 showed the same malfunction.

In CL5915 and LA3120, flowers were malformed: The petals were reflexed and the anthers not connate.

Stigma exsertion was observed in CLN1621L, LA3120, LA3320, LA2662 and CL5915. In CL5915, the stigmata protrude the sepals and petals in the bud stage already caused by obviously curtate sepals and petals. Since no comparisons with flowers developed under optimum temperatures were possible, the reason for the different stigma levels with regard to the anther cones could not be detected in the other lines.

After 9 weeks, only CLN1621L kept on flowering.

In general, the two plants of the line CLN2001A developed very inhomogeneous. While one of the plants had low numbers of opened flowers, no fruit set at all, normal developed sepals but short petals, anthers and stigmata the second showed vigorous flourishing, fruit set and no stigma exsertion.

LA3320 and LA3120 did not show any fruit set and set of a single fruit, respectively. One fruit of LA2662 and two fruits of CLN2418A showed symptoms of blossom end-rot (BER).

The pollen amount did not correlate with the number of fruits ($r^2= 0.29$; $p= 0.58$) or the percentage of fertilized fruits ($r^2= 0.47$; $p= 0.35$).

The genotypes differed in the number of fruits, the percentages of fertilized fruits and fruits containing degraded placenta tissue (colloidal tissue, Table 16), whereof, however, only the difference in fruits containing colloidal tissue was statistically significant.

The percentages of fertilized fruits and fruits with colloidal tissue correlated ($r^2= 0.84$; $p= 0.0375$).

Table 16: Average number of fruits per inflorescence (Frav Inf^{1}), sum of fruits per plant (Frsum), and percentages of fertilized fruits (Frset) and fruits with colloidal tissue (Frplac) of heat tolerant tomato genotypes from AVRDC (all accession names starting with CL of CLN) and TGRC (accession LA2662) grown under heat stress (32/ 28 °C) for 10 weeks in a climate chamber at Hannover University. Different letters within columns indicate significant differences between genotypes (SNK test, α < 0.05, n= 24).

Genotype	Frav Inf^{1}	Frsum	Frset [%]	Frplac [%]
CLN1621L	1.8 a	9 a	80.0 a	80.0 a
CLN2001A	2.3 a	7 a	91.7 a	83.3 a
CLN2418A	2.3 a	7 a	44.4 a	88.9 a
CL5915	1.5 a	12 a	66.7 a	95.8 a
FMTT260	2.5 a	5 a	0.0 a	0.0 b
LA2662	1.3 a	4 a	33.3 a	33.3 ab

Only nine varieties developed any pollen and were thus applicable for investigations. The differences between the genotypes were found to be highly significant (Table 17). The pollen release of line CLN1621L was highest amongst other lines from the AVRDC. Significant differences were found between these lines.

The pollen release was very low and none of the varieties had pollen shed high enough to conduct the staining procedure.

Table 17: Average number of pollen per flower of different heat tolerant tomato genotypes grown under heat stress for 10 weeks measured during the experiment (average of four weeks). Different letters within columns indicate significant differences between genotypes (SNK test, α < 0.05, n= 165).

Genotype	Average number of pollen	
CLN1621L	4443.9	a
CLN2001A	576.2	ab
CLN2418A	1051.5	ab
CL5915	81.9	bc
FMTT260	12.9	c
LA2662	8.0	c
LA3320	157.2	bc
LA3120	61.2	bc
LA2661	45.6	bc

The pollen release was significantly different at different sampling dates (Table 18). It diminished during the experimental course from cw12/2005 to cw14/2005 and started to increase subsequently. However, this trend was not significant (Table 18).

Table 18: Average number of pollen per flower per week of different heat tolerant tomato genotypes grown under heat stress for 10 weeks (calendar weeks [cw]) measured during the experiment (average of nine genotypes). Different letters within columns indicate significant differences between sampling dates (SNK test, α < 0.05, n= 165).

Sampling date [cw]	Average number of pollen	
12	1200.16	a
13	116.08	b
14	19.45	c
15	59.74	bc

The pollen amount declined from cw12 to cw14 in all varieties by more than 95 % except in the line CLN2001A (60 %, Table 19).

Table 19: Maximum reduction of the pollen amount per flower in nine heat tolerant genotypes grown under heat stress (32/ 28 °C) in a climate chamber for ten weeks.

Genotypes	Reduction of pollen amount [%]
CLN1621L	96.81
CLN2001A	60.27
CLN2418A	99.95
CL5915	100.00
FMTT260	97.08
LA2661	99.56
LA2662	99.24
LA3120	99.82
LA3320	99.45

3.5 Comparisons of different methods to evaluate pollen viability and pollen tube growth (*in vivo* and *in vitro*)

3.5.1 Evaluation of different pollen growth media and their comparison to FDA staining

Since not only pollen viability but pollen tube growth is important for a successful fertilization different liquid pollen growth media were used to score the percentage of pollen germination *in vitro*: the growth medium according to Brewbaker et al. (1963) (herein after referred to as 'B&K'), the B&K medium modified by Poulton et al. (2001) ('P') and a mixture of sucrose solution and boric acid.

Pollen of single flowers developed under optimum temperatures of different genotypes were harvested immediately after anthesis and 2-3 days after anthesis. They were mixed with 300 µl of one of the growth media. A drop of these pollen-suspensions was transferred to a microscope slide, covered with a cover slip, and the pollen tube growth checked visually under the microscope after laps of time of 0.5, 1, 2 and 4 hours. Pollen tubes were rated as developed when they reached at least the length of the pollen grain diameter.

Results

The experiment was repeated with the B&K medium and the P medium. Additionally the pollen tube growth in the B&K medium modified by Heslop-Harrison (1984) was investigated.

Pollen drenched with the B&K medium did not form any pollen tubes within the first two hours regardless the age of the flower. After three hours, the pollen sampled from flowers just after anthesis showed some pollen tubes. When pollen were incubated with the P medium growth of pollen tubes became visible after 0.5 hours already. During the course of the experiment, the tubes kept on growing whereas no new tubes could be found after the experiment was ceased. When pollen of three flowers were immersed in sucrose/ boric acid solution only pollen from one flower which was 2-3 days old developed pollen tubes after 1.5 hours but no further growth was observed afterwards. Pollen from the other two flowers did not develop pollen tubes at all. The pollen grains looked shriveled and sere after four hours of treatment. Because of the insufficient number of developed pollen tubes no meaningful statistical analysis could be accomplished.

In the repetition, pollen of three flowers treated with the B&K and HH media formed pollen tubes after 0.5 hours already. Only for a single pollen grain incubated with the P medium a pollen tube became visible. After 1.5 hours, the number of pollen with tubes of these samples increased in the B&K and HH medium. Additionally, tubes from pollen grains of another flower started growing. No differences between the first two media were visible.

After three hours, pollen from five out of six flowers mixed with the B&K medium developed pollen tubes while this was true only for pollen from four and three flowers mixed with the HH and P medium, respectively. In contrast to the B&K and HH media where the number of tubes increased slightly over time, no further pollen tube growth was observed in the P medium. The pollen tube growth in the HH medium was not sufficient for statistical analysis.

In the third experiment only pollen from three just opened flowers of different genotypes were examined. Pollen were harvested and mixed with 300 µl B&K or P medium, respectively. Additionally the complete anthers of three flowers of two different genotypes have been ground with 300 µl growth media.

The 3factorial ANOVA for the remaining data did not reveal any effect of the factors genotype, medium type or experimental duration (Table 20 to Table 22). The percentage of germinated pollen grains ranged from 0 % to 43 % in the P medium and from 2 % to around 71 % in the B&K medium regardless the different genotypes and the laps of time. Since no significant differences between individual experiments were found the data of all experiments were pooled.

Table 20: Average percentages of germinated pollen of three different tomato varieties (Pannovy, Typhoon, Hillmar Hellfrucht ['HH']). Means within columns followed by the same letter are not significantly different between varieties (SNK test, α< 0.05, n= 54).

Variety	Germinated pollen [%]	
Pannovy	29.96	a
Typhoon	26.82	a
HH	13.95	a

Table 21: Average percentages of germinated tomato pollen grown on two different growth media ((Brewbaker et al., 1963)'B&K'(Poulton et al., 2001), 'P'). Means within columns followed by the same letter are not significantly different between pollen growth media (SNK test, α < 0.05, n = 54).

Medium	Germinated pollen [%]	
B&K	22.24	a
P	22.88	a

Table 22: Average percentages of germinated tomato pollen grown on growth media for 1, 2 or 4 hours (h). Means within columns followed by the same letter are not significantly different between growth durations (SNK test, α < 0.05, n = 54).

Growth duration [h]	Germinated pollen [%]	
1	26.56	a
2	25.85	a
4	24.27	a

Furthermore a comparison between pollen germinated in the P medium and the pollen viability obtained by FDA staining was done. Pollen were harvested immediately after anthesis and divided into two subsamples of similar pollen quantity. One half was drenched with 300 µl of the P medium and the pollen tube growth evaluated after one hour. The second half was stained with FDA and evaluated under UV light.

While the mean percentage of pollen germination in the growth media was 6 %, the mean of the pollen viability evaluated via FDA was 58 % and the data did not correlate (r^2= 0.22, p= 0.4643).

3.5.2 Aniline blue staining versus FDA staining

For the evaluation of the pollen tube growth *in vivo* gynoecia were stained with aniline blue. Aniline blue binds to the glucose units building the callose polymer, a structural component of cell walls of pollen tubes. Since the style tissue does not contain callose aniline blue differentiates well between these two structures. The following results were obtained in the context of a Bachelor thesis (Schmidt, 2007).

Flowers were pollinated by hand and harvested after 24 hours. The sepals, petals and anthers were removed from the flowers and the gynoecia were entirely covered with a fixer in a glass tube for 24 hours.

The fixer was removed by washing with DI and the plant tissue was macerated in 1 M sodium hydroxide at 60 °C for 45 min. After removing the sodium hydroxide solution with DI the samples were stained with the aniline blue solution for 15 min. The gynoecium was transferred to a microscope slide and covered with glycerin and a cover slip to avoid early desiccation. The preparation was squeezed cautiously under the cover glass. Leaked glycerin was carefully removed with a tissue. The total number of pollen and the pollen tubes in the style which were grown into the ovules were rated visually under the microscope.

As displayed in Figure 13 the amount of pollen on the stigmatic surface was very high and it was impossible to rate the number of pollen precisely and to follow the single pollen tubes through the style tissue down to the ovules. The pollen tubes grew bundled in the style channel before they separated again in the ovary sac. Since the vascular bundles showed the same coloration than the pollen tubes it was difficult to distinguish between these structures.

Therefore, a quantitative analysis of the observations was impossible and the interpretation could be done on a qualitative basis only. The comparison of the FDA staining method and the aniline blue staining method showed a correlation: the higher the percentage of viable pollen (estimated by FDA staining) the more pollen tubes were found to grow through the style (estimated by aniline blue staining). Contrarily the more non-viable pollen were found via FDA staining the less pollen tubes could be detected via aniline blue staining.

The numbers of pollen tubes growing from the style into the ovaries and entering the ovules were estimated of approximately five to twenty.

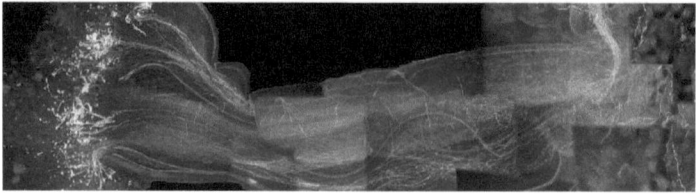

Figure 13: A gynoecium (50fold magnification) of a tomato flower stained with aniline blue: the stigma (red on the left side) and the attached style down to the ovary containing many ovules (brown round shaped on the right side). The pollen grains are visible as small blue-green dots on the stigma and the pollen tubes as thin blue-green lines through the style. (The picture is composed of several individual photographs mounted using an image processing software)

3.5.3   MTT staining versus FDA staining

For the reduction of costs and technical equipment a method was tested for which no UV-light for the determination of pollen viability is required. Yellow MTT gets reduced to purple formazan in the mitochondria of living cells (Figure 14). This reduction takes place only when mitochondrial reductase enzymes are active, and therefore conversion can be directly related to the number of viable (living) cells.

In order to asses the applicability of the staining method using tetrazolium bromide (MTT) instead of FDA pair wise regressions with each one of the three classes of viable, slightly degraded, and non-viable pollen stained with either of the two staining methods were conducted (Figure 15 to Figure 17).

Pollen of single flowers were harvested immediately after anthesis and the samples were divided into two subsamples of similar pollen amounts. One half each was stained with FDA or MTT. Both procedures were undertaken as contemporary as possible.

Pollen were placed on a microscope slide with a pipette tip and a single drop of staining solution was added. The color of at least 100 pollen grains was rated after five minutes incubation time at room temperature. Pollen were rated as viable or dead when showing dark violet or no or coloration, respectively. Degraded pollen showed a bright violet coloration.

The Pearson correlation coefficients (r) were 0.76 ($p< 0.001$), 0.60 ($p< 0.001$), and 0.84 ($p< 0.001$) for the classes viable pollen, degraded pollen, and non-viable pollen, respectively. The coefficients of determination (r^2) for these classes were 0.87 ($p< 0.001$), 0.77 ($p< 0.001$), and 0.92 ($p< 0.001$).

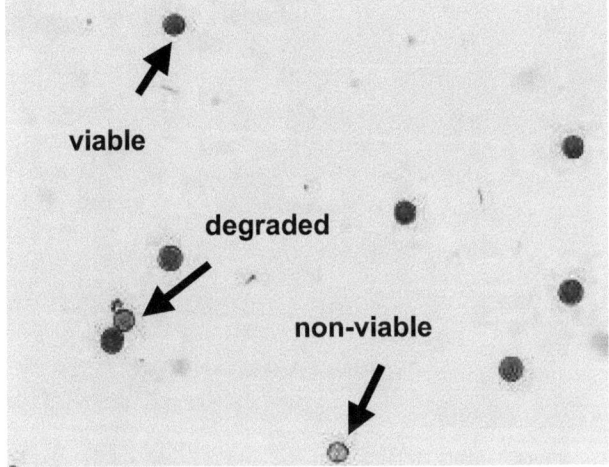

Figure 14: Pollen stained with MTT stain in different extensities depending on their viability. Viable pollen appear dark violet, degraded pollen less dark and non-viable pollen do not stain at all. The picture was taken under the microscope with 100fold magnification.

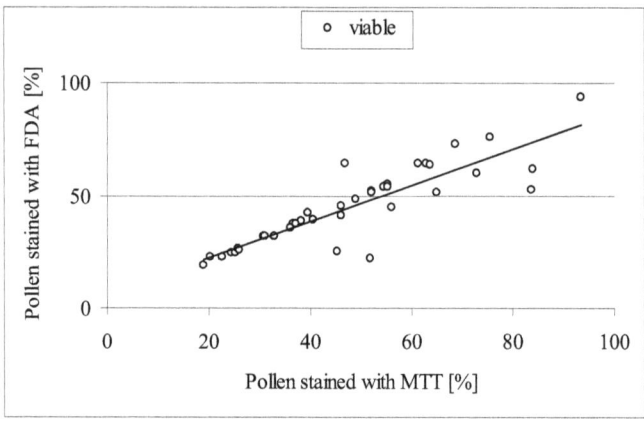

Figure 15: Correlation between the percentages of pollen rated as viable stained either using the FDA or the MTT method ($r^2= 0.87$, $p< 0.001$, $n= 38$).

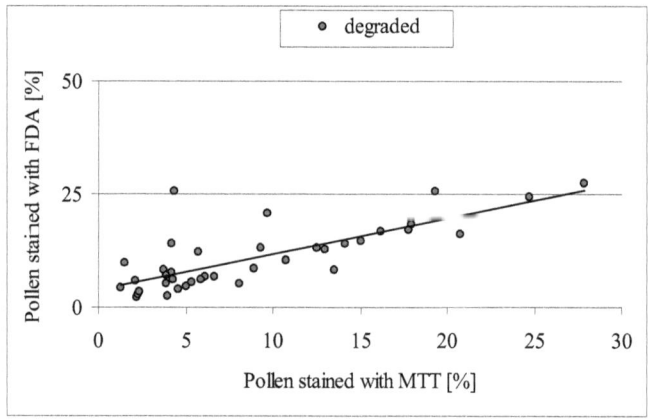

Figure 16: Correlation between the percentages of pollen rated as degraded stained either using the FDA or the MTT method ($r^2= 0.77$, $p< 0.001$, $n= 38$).

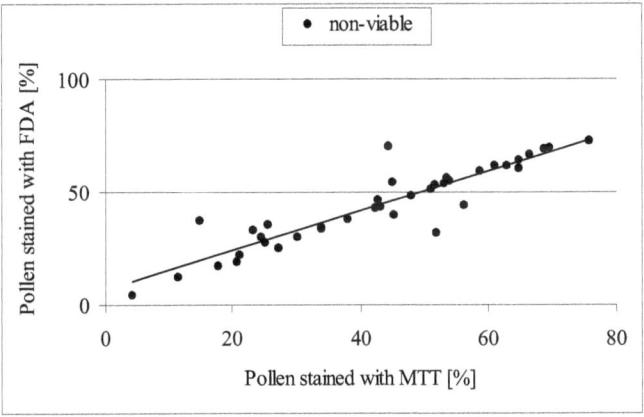

Figure 17: Correlation between the percentages of pollen rated as non-viable stained either using the FDA or the MTT method (r^2= 0.92, p< 0.001, n= 38).

3.6 Pollen storage

Since the number of pollen samples harvested per sampling date was too high for immediate processing, but comparability had to be assured a series of experiments was conducted to develop an applicable procedure for the storage of pollen samples. This storage possibility was of particular importance for experiments with more than 20 samples per harvesting date because the pollen counting and staining procedures were laborious and not to handle on the same day.

Pollen of the line CL5915 grown at different times under diverse conditions were harvested and stored under different conditions. Pollen of five individual flowers of different plants were harvested at the same time and each sample was divided. One half each was investigated immediately (day0) and the second half stored for diverse laps of time. The change of the pollen viability of each sample was measured using the sample investigated immediately as 100 % standard. The stored samples were cooled down to 4 °C and stored in the staining solution at room temperature, respectively. After 1, 3, 9, and 38 days (d), their viability was determined via FDA staining. The experiment was repeated four times.

Duration of the pollen storage significantly influenced the viability of the pollen. The results for day 1 and 3 did not differ but a significant loss of viability was found after 9 d of storage at 4 °C (Figure 18). The increment of degraded and non-viable pollen was not significantly influenced by the storage duration (degraded pollen: p= 0.06; non-viable pollen: p= 0.25).

Samples stored in the staining solution showed a loss of viability of 100 % after one day already. Therefore, no sufficient statistical analysis was possible.

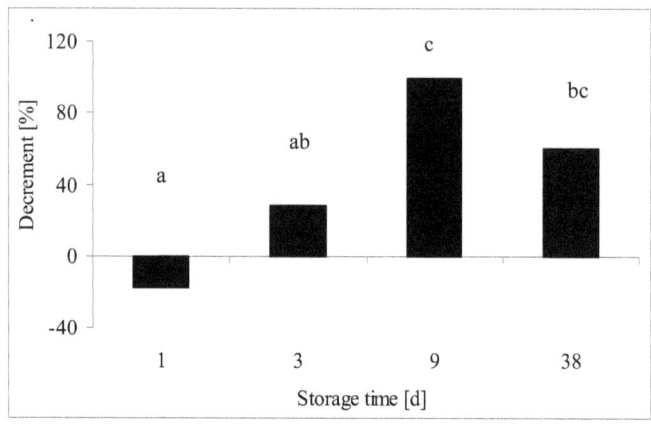

Figure 18: Changes in the average proportion of viable pollen of the tomato line CL5915, sampled in different greenhouses at different times and stored at 4 °C for different time spans. Means with different letters are significantly different (Tukey test, α< 0.05, n= 16).

3.7 Genetic variability in heat tolerant tomato lines under greenhouse conditions in Thailand

The genetic variability of 16 known or supposed to be heat tolerant tomato lines was evaluated in a GH situated in Thailand during the rainy season from August to September 2005 under natural light conditions.

The irrigation frequency depended on solar radiation integral and the duration of the dripper intervals were regularly adjusted according to the plant age. The air temperature and relative humidity inside and outside the GHs were monitored continuously with the aid of aspirated psychrometers (sensors: sheathed type K [NiCr-Ni] thermocouples, diameter 0.5 mm) (BGT, Hannover, Germany). Every 15 sec data were measured, transferred to a purpose-build data-logging system (BGT, Hannover, Germany), and stored as mean values every 5 min. All sensors were calibrated before the experiment started.

Five plants of the genotypes 1- 16 (80 plants in total) listed in Table 23 were transplanted from the evaporative cooled nursery to the experimental GH two weeks after sowing in cw30/2005. Additional 144 plants of different introgression lines (ILs) and 136 plants of the hybrid FMTT260 were added to fill in the gaps and secure a consistent plant density. Plants were placed in six rows each with 60 plants per row. The plants were completely randomized within the four central rows. The gaps were filled with FMTT260. Five plants of FMTT260 in the middle rows were randomly chosen as experimental plants and labeled. The side rows consisted of FMTT260 plants only. Plants of indeterminate cultivars were cultivated with one shoot and pruning was done twice a week. In case of determinate genotypes no pruning was done. Senescent leaves were removed regularly up to the first fruit-carrying truss and plants were laid down according to necessity.

Table 23: 16 genotypes known or suspected to be heat tolerant used for the screening of heat tolerance under Thailand greenhouse conditions. All genotypes were grown in the rainy season in 2005 for 10 weeks.

Number	Variety/ line
1	ChiaTai
2	CL5915
3	CLN1621L
4	CLN2001A
5	CLN2418A
6	Donna091
7	FMTT260
8	FMTT269
9	HT7
10	LA2661
11	LA2662
12	LA3120
13	LA3320
14	Pannovy
15	Sida013
16	Valentine

Results

The collection of all plant related data was restricted to the plants of the four central rows in order to eliminate any possible interferences resulting from the plant's position close to the sidewalls. The evaluation of the seed set was recorded once at the end of the experiment by cutting the fruits. The data collection was commenced in cw32/2005.

The weekly increase in plant height and the maximum plant height did not significantly differ within one line or between different lines of one growth habit (determinate/ indeterminate). In general, the weekly increase of the height of determinate growing genotypes was less than that of indeterminate growing plants and ranged from 17 centimeters in line LA2662 to 40 centimeters in variety Pannovy resulting in 78 and 156 cm total plant heights for the respective cultivars at the end of the experiment.

The number of flowers, the total number of fruits, and the number of parthenocarpic fruits per inflorescence differed significantly between the lines/ varieties (Table 24)

The lowest total number of flowers and number of flowers per inflorescence were observed in Pannovy while the highest total number of flowers was found in Valentine. The number of flowers per inflorescence was highest in FMTT269 (Table 24). In general, higher numbers of flowers were observed in determinate growing cultivars as compared to indeterminate growing types.

The lowest amount of fruits and number of fruits per inflorescences were observed in LA3120 and the highest values were reached by Donna and LA2662. The lowest fruit set was found in LA3120 and the highest in Pannovy (Table 24).

A difference of fruit set between plants depending to their position in the GH was detectable. Plants close to the entrance- and therefore close to the fans- and at the opposite end of the GH showed a higher fruit set compared to plants placed in the center of the GH. The lowest sum of parthenocarpic fruits were obtained by LA3120 and the highest by Donna, while the lowest and highest number of parthenocarpic fruits per inflorescence were found in CLN2001A and LA2662, respectively (Table 24).

The genotypes differed significantly in their pollen viability: The highest percentage of viable pollen and lowest percentage of non-viable pollen were produced by CLN1621L, the lowest number of viable pollen and highest number of non-viable pollen by FMTT260. The number of slightly degraded pollen did not significantly differ between the varieties (Table 24).

Table 24: Average weekly height increase (growth, [cm week^{-1}]), sum of flowers (Flsum), average (Flav Inf^{-1}) and maximum (Flmax Inf^{-1}) number of flowers per inflorescence, sum of fruits (Frsum), average (Frav Inf^{-1}) and maximum (Frmax Inf^{-1}) number of fruits per inflorescence, average percentages of parthenocarpic fruits (Frparth), viable pollen (Poll), pollen amount per flower (Poll Fl^{-1}) and set fruits (Frset) of 16 tomato genotypes known or supposed to be heat tolerant evaluated in greenhouses in the rainy season 2005 in Central Thailand. Means within columns followed by the same letter are not significantly different between genotypes (SNK test, α< 0.05, n= 54).

Genotype	Growth	Flsum	Flav Inf^{-1}	Flmax Inf^{-1}	Frav Inf^{-1}	Frsum	Frmax Inf^{-1}	Frparth [%]	Poll [%]	Poll Fl^{-1}	Frset [%]
ChiaTai	28.5 a	2434 a	8.8 cd	33 ab	1.3 cde	362 a	13 a	99.4 a	.	1.0 ef	14.8 c
CL5915	21.9 a	2325 a	9.6 cb	30 ab	1.3 cde	324 a	15 a	36.4 ef	37.1 ab	18745.7 a	13.0 cd
CLN1621L	34.1 a	2731 a	7.9 ed	25 ab	2.2 ab	753 a	18 a	63.1 cd	40.7 a	3426.3 abc	29.1 b
CLN2001A	21.8 a	2726 a	11 b	39 a	1.1 def	291 a	11 a	29.7 f	.	9858.9 ab	10.6 cd
CLN2418A	34.1 ab	1336 a	8.9 cd	31 ab	0.9 ef	134 a	11 a	50.2 de	8 c	359.7 e	12.2 cd
Donna	34.1 a	2692 a	6.3 f	24 b	2.0 bc	837 a	21 a	89.6 a	16.4 c	1052.4 abcd	26.6 b
FMTT260	38.9 a	554 b	9.7 bc	28 ab	2.4 ab	135 a	12 a	71.7 bc	2.4 bc	484.0 e	31.9 b
FMTT269	36.6 a	801 b	12 a	25 ab	1.7 bcd	118 a	19 a	58.6 cd	.	11.0 def	15.9 c
HT7	26.3 a	885 b	6.6 ef	22 b	2.1 bc	278 a	20 a	96.5 a	.	3.8 ef	28.8 b
LA2661	28.1 a	2352 a	5.9 f	34 ab	2.0 bc	781 a	24 a	95.6 a	7.2 c	30.4 cdef	30.5 b
LA2662	17.1 a	788 b	9.7 bc	23 ab	2.8 a	230 a	15 a	73.0 bc	.	11.3 def	25.1 b
LA3120	20.9 a	2609 a	7.9 de	25 ab	0.1 g	45 a	6 a	85.0 ab	.	5.6 ef	1.4 e

LA3320	24.8 a	2795 a	8.2 cd	25 ab	0.5 ef	306 a	13 a	98.7 a	. .	0.8 ef	7.5 cde
Pannovy	39.9 a	511 b	5.7 f	14 b	2.2 ab	195 a	12 a	98.2 a	0.8 c	232.5 e abcd	39.1 a
Sida	30.1 a	2829 a	9.2 cd	31 ab	0.7 efg	227 a	14 a	85.0 ab	5.7 ab	79.7 f bcde	7.4 cde
Valentine	31.1 a	3033 a	7.9 de	30 ab	0.4 fg	158 a	8 a	50.2 de	9.0 c ab	2268.6 abc	5.2 de

Results

In pair wise comparisons, the lines CL5915, CLN2001A, and CLN1621L supplied by the AVRDC performed better compared to the standard variety FMTT260 (Table 25). Only in line CL5915 a significantly higher mean number of viable pollen in a pair wise comparison with the negative control Pannovy was found though all lines showed positive mean differences (Table 26).

Table 25: Differences in average pollen amounts between heat tolerant lines supplied by the AVRDC, Taiwan, and variety FMTT260 (AVRDC, Taiwan). Significant differences are labeled with *** and not significant differences with n.s.

Genotype			Differences between means
CL5915	FMTT260	3.66	***
CLN2001A	FMTT260	3.01	***
CLN1621L	FMTT260	1.96	n.s.
CLN2418A	FMTT260	-0.30	n.s.
Pannovy	FMTT260	-0.73	n.s.

Table 26: Differences in average pollen amounts between heat tolerant lines supplied by the AVRDC, Taiwan, and with the negative control Pannovy (Syngenta Seeds, Germany) evaluated in a greenhouse in the rainy season 2005 in Central Thailand. Significant differences are labeled with *** and not significant differences with n.s.

Genotype			Differences between means
CL5915	Pannovy	4.39	***
CLN2001A	Pannovy	3.75	n.s.
CLN1621L	Pannovy	2.69	n.s.
CLN2418A	Pannovy	0.73	n.s.
FMTT260	Pannovy	0.44	n.s.

The pollen viability of different lines reacted differently in the same time period. While the pollen viability increased in line CLN2418A, Donna and CLN2001A within the first two weeks it decreased in line CLN1621L or remained equal in line CL5915 and variety FMTT260 (Figure 19 to Figure 24). Also in later periods the pollen viability of the different lines did not react equally.

No correlation between the development of the pollen viability (Figure 19 to Figure 24) and the development of an individual external factor such as relative air humidity, air or soil temperature (Figure 25 and Figure 26) was observable.

Percentages of viable pollen of the CLN1621L were most stable and ranged from 60 to 70 % within the duration of the experiment. Results in cw36/2005 (sampling date 05.09.2005) were an exception. In this week, in all genotypes a severe drop of the percentage of viable pollen was observed (Figure 19 to Figure 24). The percentages decreased to zero when they showed any vital pollen before except in lines CLN2001L and CLN1621L. The total loss of pollen viability in line CLN2001L was found one week earlier (sampling date 31.8., Figure 23) while the pollen viability of line CLN1621L did not drop to less than 16 % (Figure 24).

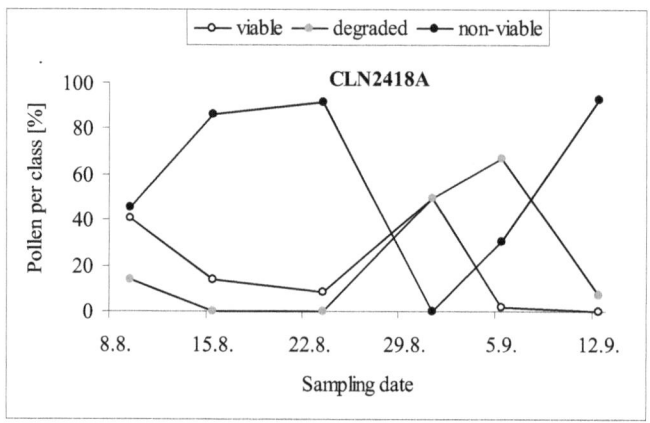

Figure 19: Development of the percentages of viable pollen, degraded pollen, and non-viable pollen for line CLN2418A during the course of a greenhouse experiment in Central Thailand in the rainy season 2005.

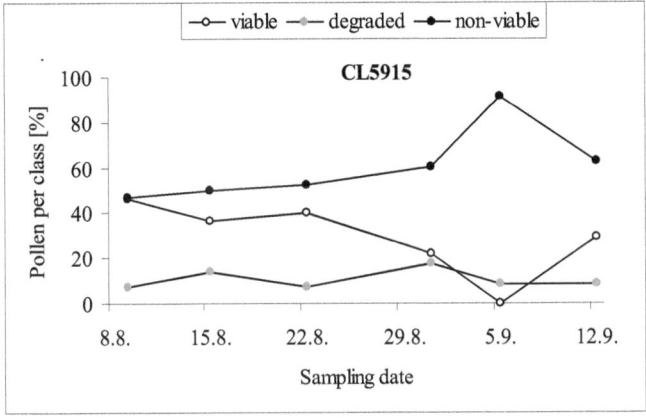

Figure 20: Development of the percentages of viable pollen, degraded pollen, and non-viable pollen for line CL5915 during the course of a greenhouse experiment in Central Thailand in the rainy season 2005.

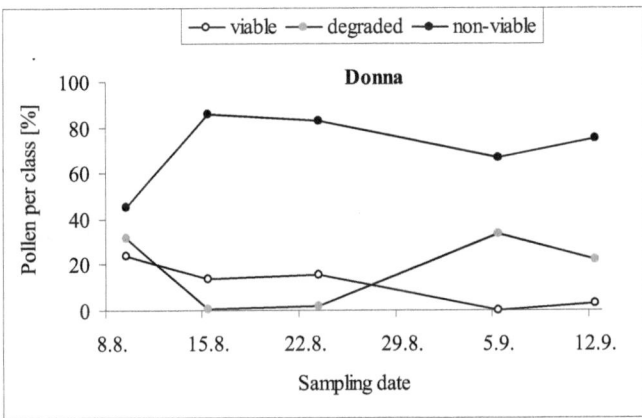

Figure 21: Development of the percentages of viable pollen, degraded pollen, and non-viable pollen for line Donna during the course of a greenhouse experiment in Central Thailand in the rainy season 2005.

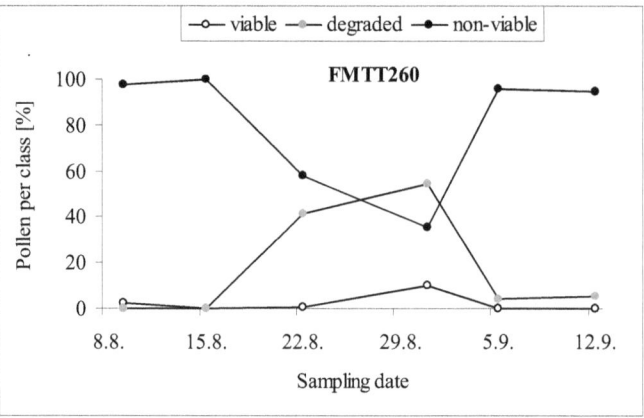

Figure 22: Development of the percentages of viable pollen, degraded pollen, and non-viable pollen for line FMTT260 during the course of a greenhouse experiment in Central Thailand in the rainy season 2005.

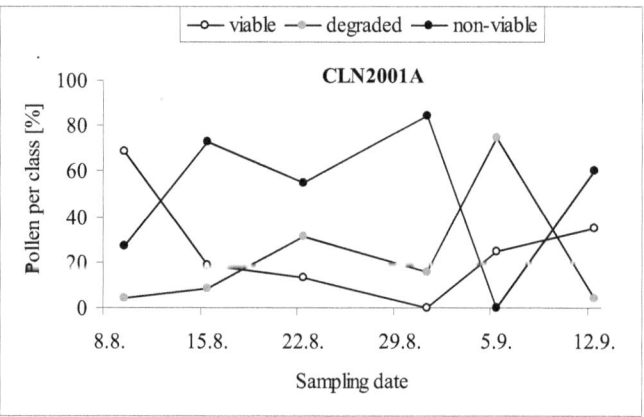

Figure 23: Development of the percentages of viable pollen, degraded pollen, and non-viable pollen for line CLN2001A during the course of a greenhouse experiment in Central Thailand in the rainy season 2005.

Results

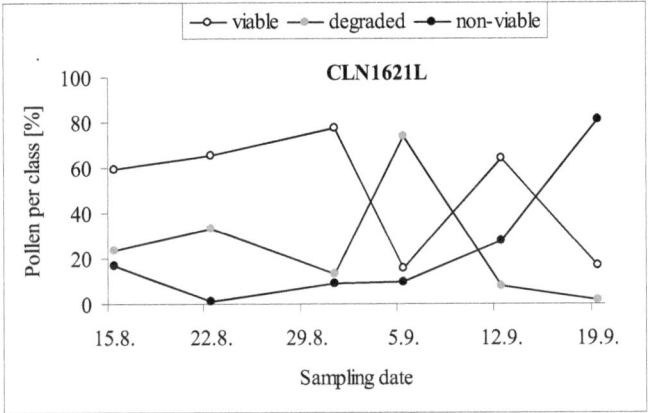

Figure 24: Development of the percentages of viable pollen, degraded pollen, and non-viable pollen for line CLN1621L during the course of a greenhouse experiment in Central Thailand in the rainy season 2005.

Figure 25 and Figure 26 show the profiles of relative air humidity, air temperature, and substrate temperature for the duration of the experiment. The curve progressions of substrate and air temperature showed similar patterns. With increasing air temperature the substrate temperature increased as well though the differences between both were more pronounced in the beginning of the experiment. From 26.08.2005 to the end of the experiment the values were almost similar.

One day before the pollen viability broke down (04.09.2005) the air humidity reached 96 % which was the highest value in the experimental progress so far (Figure 25) but it reached the same value again on 09.09.2005 and 14.09.2005 with no subsequent break down in pollen viability. At the same day the air temperature was also very high (31 °C) but it reached the same temperatures two times before on 03.08.2005 and 26.08.2005 with no subsequently complete loss of pollen viability. The substrate temperature in contrast reached its maximum temperature the first and only time during the experiment and exceeded 32 °C (Figure 26) on 04.09.2005.

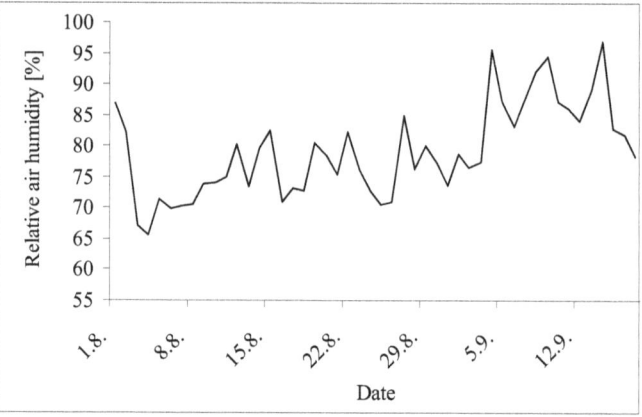

Figure 25: Profile of the relative air humidity inside the experimental greenhouse in Central Thailand during the rainy season 2005.

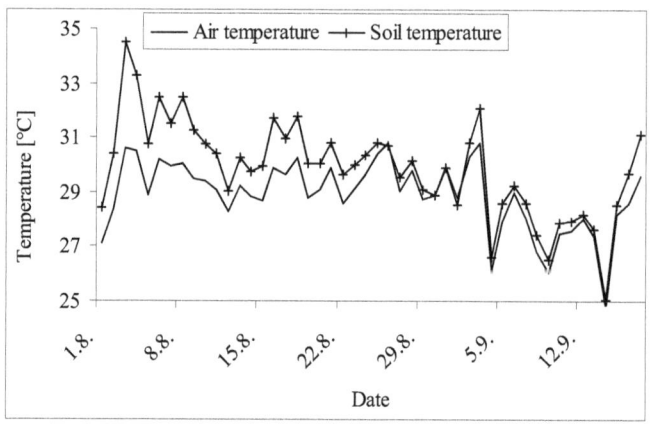

Figure 26: Profiles of air and substrate temperatures inside the experimental greenhouse in Central Thailand during the rainy season 2005.

The fruit set correlated with the sum of flowers and with the number of flowers per inflorescence at significance levels of 0.05 and 0.1, respectively (Table 27).

Both, the pollen amount and the pollen viability did not correlate with the sum of fruits or the number of fruits per inflorescence (Table 27). Neither correlated the percentage of viable pollen with the percentage of parthenocarpic fruits. But the pollen amount per flower correlated with the mean number of parthenocarpic fruits and the percentage of parthenocarpic fruits (Table 27).

Table 27: The Pearson correlation coefficients and significances for the traits flowers (Flav Inf^{-1}), fruits (Frav Inf^{-1}) and parthenocarpic fruits per inflorescence (Fr parth Inf^{-1}), sum of flowers (Flsum) and fruits (Frsum), percentages of parthenocarpic fruits (Fr parth), fruit set (Frset), and viable pollen (Poll), and the pollen amount per flower (Poll Fl^{-1}). Significances at levels of 0.05 are indicated by **, at levels of 0.1 are indicated by *, and not significant results are depicted by n.s.

	Flav Inf^{-1}	Frav Inf^{-1}	Flsum	Frsum	Frparth Inf^{-1}	Frparth [%]	Frset [%]	Poll [%]	Poll Fl^{-1}
Flav Inf^{-1}		0.11 n.s.					0.44 *		
Frav Inf^{-1}								0.10 n.s.	0.02 n.s.
Flsum				0.13 n.s.			0.59 **		0.19 n.s.
Frsum								0.45 n.s.	0.20 n.s.
Frparth Inf^{-1}								0.32 n.s.	0.70 **
Frparth [%]								0.53 n.s.	0.73 **
Frset [%]									
Poll [%]									
Poll Fl^{-1}									

3.8 Introgression lines

48 *S. pennellii* introgression lines sown in cw24/2005 and transplanted to the GH in cw27/2005 were investigated under GH growth conditions in Thailand in the rainy season to evaluate their response on heat stress. The plant data were evaluated 5 times from cw30/2005 to cw36/2005 for all plants.

From the 48 introgression lines only 37 could be evaluated while the other eleven lines had to be removed from the GH because of virus infections. The remaining lines differed significantly regarding all traits evaluated: the number of inflorescences, the number of flowers, fruits, and parthenocarpic fruits per inflorescence, the percentages of parthenocarpic fruits and fruit set (Table 28).

The number of inflorescences ranged from 4 in line 5-4 to 91 in line 10-1. These two lines were significantly worse or better, respectively, as compared to all other lines.

The mean of flowers per inflorescence varied from 2 in line 3-1 to 12 in line 2-2. These two lines performed significantly worse or better, respectively, compared to the other lines.

The mean number of fruits per inflorescence reached from 0.1 in line 3-5 and 5-4, respectively, to 3.1 in line 3-1. The lines 3-1, 1-3, and 12-3 did not differ from each other but from all other lines.

The percentage of parthenocarpic fruits ranged from 20 % in line 2-4 to 100 % in 13 different lines. The lines 2-4 and 5-4 did not differ significantly from each other. Their percentage of parthenocarpic fruits was 20 and 25 %. The next better performing line was line 2-3 with a percentage of 41 %.

The fruit set ranged from 21 % in line 8-1 to 72 % in line 1-3. For this parameter, no single line performed best or worst.

The weekly increment of the plant height ranged from 2.7 cm in line 5-5 to 45 cm in line 6-3 but did not differ significantly. Only line 2-5 showed a determinate growth within the experimental duration.

Results

Table 28: Average number of flowers (Flav Inf^{-1}) and fruits per inflorescence (Frav Inf^{-1}), percentage of parthenocarpic fruits (Frparth) and set fruits (Frset) of 37 tomato introgression lines grown in a greenhouse during the rainy season 2005 in Central Thailand. Different letters within columns indicate significant differences between genotypes (SNK test, α< 0.05, for flowers and fruits per inflorescence: n= 5273; for percentages of parthenocarpic fruits and fruit set: n= 1810).

Genotype	Flav Inf^{-1}		Frav Inf^{-1}		Frparth [%]		Frset [%]	
IL1-3	5.56	hijk	2.94	a	100.00	a	72.48	a
IL1-4	6.96	cdefghi	1.05	bcde	96.88	a	46.08	bcdefg
IL2-2	11.82	a	1.22	bcde	100.00	a	33.00	cdefg
IL2-3	6.64	efghij	0.60	bcde	40.71	c	35.57	cdefg
IL2-4	7.70	bcdefgh	0.88	bcde	20.42	d	48.68	abcdef
IL2-5	8.58	bcdef	1.93	b	95.88	a	40.11	cdefg
IL2-6	5.52	hijk	1.73	bcd	100.00	a	60.04	abc
IL3-1	9.16	bc	3.10	a	100.00	a	49.44	abcdef
IL3-2	6.22	ghijk	0.37	de	100.00	a	41.48	cdefg
IL3-4	11.47	a	1.73	bcd	100.00	a	36.92	cdefg
IL3-5	4.61	jk	0.15	e	100.00	a	33.00	cdefg
IL4-3	6.93	cdefghi	0.86	bcde	87.82	ab	38.54	cdefg
IL4-4	7.94	bcdefg	1.94	b	96.67	a	51.03	abcde
IL5-1	6.15	ghijk	1.12	bcde	100.00	a	41.63	cdefg
IL5-2	8.88	bcde	0.78	bcde	97.00	a	35.72	cdefg
IL5-3	9.74	b	0.56	bcde	100.00	a	46.73	bcdefg
IL5-4	2.30	l	0.40	de	25.00	d	48.61	abcdef
IL5-5	9.59	b	0.86	bcde	79.43	b	22.51	fg
IL6-4	6.17	ghijk	0.36	de	100.00	a	39.46	cdefg
IL7-1	6.84	defghi	1.59	bcd	98.00	a	50.53	abcde
IL7-2	8.95	bcd	1.34	bcde	96.96	a	37.39	cdefg
IL8-1	8.22	bcdefg	0.58	bcde	100.00	a	20.66	g
IL8-2	7.10	cdefghi	1.01	bcde	97.21	a	47.67	abcdefg
IL9-1	4.39	k	1.38	bcde	100.00	a	67.69	ab
IL9-2	9.79	b	1.83	bc	100.00	a	40.34	cdefg
IL9-3	7.72	bcdefgh	0.55	bcde	80.23	b	26.93	efg
IL10-1	5.27	ijk	0.80	bcde	100.00	a	52.27	abcde
IL10-2	9.18	bc	1.34	bcde	100.00	a	22.16	fg
IL10-3	7.96	bcdefg	1.11	bcde	100.00	a	47.80	abcdefg

Results

IL11-1	6.64	efghij	0.51	cde	100.00	a	37.40	cdefg
IL11-2	6.37	fghijk	0.42	de	100.00	a	42.56	bcdefg
IL11-3	9.15	bc	1.05	bcde	98.04	a	34.81	cdefg
IL11-4	9.73	b	1.11	bcde	100.00	a	28.83	defg
IL12-1	7.75	bcdefgh	1.38	bcde	99.71	a	45.71	bcdefg
IL12-2	6.42	fghijk	1.75	bcd	100.00	a	55.70	abcd
IL12-3	7.21	cdefghi	2.90	a	99.61	a	53.74	abcde
IL12-4	7.88	bcdefg	0.62	bcde	100.00	a	33.71	cdefg

3.9 Plant response to different greenhouse set-ups

A major objective of the 'Protected Cultivation Project' was the development of suitable GH cooling methods to improve crop performance, tomato fruit yield and quality under tropical climate conditions. Therefore, various experiments were conducted in which different mesh sizes, roof covers, and active cooling methods were compared regarding their influence on the micro-climatically conditions inside the GHs and their impact on plant growth and pollen performance.

At three dates in July 2005 pollen samples were taken from three plants per GH in order to figure out whether or not the differential micro climate would exert an influence on pollen viability and amount. The roof plastics of two of the GHs were additionally coated with a NIR-reflecting pigment. The GH floors were covered with the bicolored (black/ white) plastic mulch. In two GHs, the white, in the other two the black side of the mulch was turned upside, resulting in four combinations of GH properties (Table 29)

Table 29: The four different greenhouse set-ups with different roof covers and ground mulch colors used to investigate the influence of NIR-shading paint and different mulch colors on plant growth, pollen amount, and pollen viability.

Set-up	Mulch color	Roof cover
'wref'	white	NIR-reflecting
'bref'	black	NIR-reflecting
'wtrans'	white	NIR-transmissive
'btrans'	black	NIR-transmissive

Results

In this experiment with different roof covers (NIR-reflecting or NIR-transmissive) and ground mulches (black and white surface upside), a significant influence of the mulch color and the NIR reflecting pigment on the plant height was found (Table 30). At the sampling date plants in 'bref' had grown tallest while those in 'wtrans' were shortest.

The pollen viability was significantly higher in houses with white covered grounds while the roof properties did not influence the pollen viability (Table 30) and the pollen amount did not differ between any of the roof-groundcover-combinations (Table 30). However, the trend showed a higher pollen amount in 'btrans'.

Table 30: Averages of final plant height, pollen viability (Poll) and pollen amount (Poll Fl^{-1}) of the tomato variety FMTT260 grown in greenhouses with four different set-ups in the rainy season 2005 in Central Thailand. The set-ups differed in their mulch color (black ['b'] or white ['w']) and their roof properties (NIR-transmissive ['trans'] or coated with NIR-shading paint ['ref']). Different letters within columns indicate significant differences between set-ups (pollen viability/ amount: SNK test, α< 0.05, n= 22; plant height: Tukey test, α< 0.05, n= 12).

Set-up	Plant height [cm]		Poll [%]		Poll Fl^{-1}	
'bref'	147.18	a	3.13	a	54,851.71	a
'wref'	142.64	b	28.48	b	55,330.30	a
'btrans'	127.00	c	4.59	a	83,724.35	a
'wtrans'	121.22	d	24.26	b	53,021.77	a

Based on the results of these preliminary investigations the influence of the NIR-reflecting pigment in combination with different ground mulch colors was investigated in a further experiment during the dry season from cw52/2005 to cw19/2006. Sampling of vegetative plant growth was commenced once a week from cw1/2006. Pollen were harvested twice a week and counted. The viability was evaluated by staining with MTT. The collection of pollen data commenced in cw5/2006.

The pollen viability was significantly influence by the roof cover (Table 31). The percentage of viable pollen was significantly lower in the houses with NIR-reflecting roofs compared to houses with NIR-transmissive roofs. No influence of the mulch color on pollen viability could be observed.

The percentage of viable pollen declined during the experiment but not continuously. It increased and decreased in a weekly rhythm with lower increases leading to lower percentages of viable pollen at the end of the experiment (Figure 27) remaining finally at a level of around 20 % from

Results

cw12 until the end of the experiment in cw16. The percentages of viable pollen decreased significantly in week seven. In week eight, the value rose significantly then dropped again in week nine. Another significant decline was found from week ten to twelve.

No significant influence of the different roof covers or mulch colors on the pollen amount was detectable (Table 31) though a clear tendency to higher pollen amounts was observable in houses with black ground mulch.

The pollen amount declined during the experiment but not continuously. It increased significantly for two weeks to its maximum in cw7 and decreased afterwards to less than 1000 per flower in cw14. From cw14 to cw15 the pollen amount doubled to drop again afterwards to almost zero (Table 32). While in 'bref', 'btrans', and 'wtrans' a slight increase in the pollen amount was observable in week ten, a clear enhancement was found in 'wref' one week earlier.

Table 31: Average proportion of viable pollen (Poll) and pollen amount per flower (Poll Fl^{-1}) of tomato variety FMTT260 grown in greenhouses with different set-ups during the dry season 2006 in Central Thailand. The set-ups differed in their mulch color (black ['b'] or white ['w']) and their roof properties (NIR-transmissive ['trans'] or coated with NIR-shading paint ['ref']). Different letters within columns indicate significant differences between the set-ups (SNK test, $\alpha < 0.05$, pollen viability: n= 350; pollen amount: n= 359).

Set-up	Poll [%]		Poll Fl^{-1}	
'bref'	29.01	a	6,279.72	a
'wref'	29.34	a	4,245.61	a
'btrans'	31.85	b	9,076.95	a
'wtrans'	31.54	b	5,727.17	a

Table 32: Average percentage of viable pollen (Poll) and pollen amount per flower (Poll Fl^{-1}) of tomato variety FMTT260 averaged across four greenhouse set-ups in the dry season 2006 in Central Thailand. Different letters within columns indicate significant differences between sampling dates (calendar weeks, [cw]) (SNK test, $\alpha < 0.05$, pollen viability: n= 350, pollen amount: n= 359).

Sampling date [cw]	Poll [%]		Poll Fl^{-1}	
5	52.91	a	21,816.64	b
6	55.86	a	34,207.50	b
7	26.23	cd	70,003.75	a
8	38.70	b	25,402.20	b
9	27.96	cd	16,802.40	b
10	32.61	bc	13,935.46	b
12	20.47	d	2,293.77	c
13	22.24	cd	1,782.58	cd
14	22.11	cd	835.84	de
15	20.84	d	1,529.47	cd
16	18.67	d	454.07	e

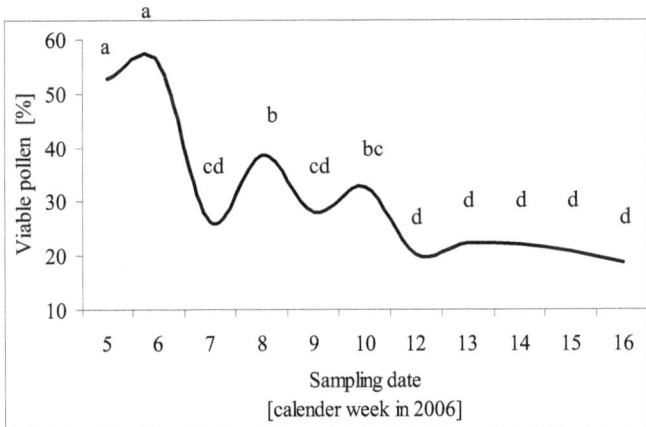

Figure 27: Development of the viability of pollen sampled from tomato plants (variety FMTT260) for 12 weeks (calendar week [cw] 5 to16) during the dry season 2006 in Central Thailand. Values are means across four different greenhouse set-ups. Means with different letters are significantly different between sampling dates (SNK test, $\alpha < 0.05$, n= 350).

Results

The number of flowers per inflorescence decreased significantly with plant age (Table 33) whereas the number of inflorescences did not differ significantly between the treatments (Table 34).

From the fourth inflorescence, the number of developed flowers per inflorescence sank continuously. The number of flowers per inflorescence ranged from eight to nineteen.

Table 33: Average number of flowers per inflorescence of inflorescences of different age (1= first developed inflorescence) of the tomato cultivar FMTT260 during the dry season 2006 in Central Thailand. Values are means across four different greenhouse set-ups. Means followed by different letters are significantly different (SNK test, $\alpha < 0.05$, n= 571).

Inflorescence number	Flowers per inflorescence	
1	10.21	bcd
2	15.71	ab
3	14.95	abc
4	18.98	a
5	15.38	ab
6	15.43	ab
7	13.64	abcd
8	11.62	bcd
9	10.15	bcd
10	9.00	bcd
11	8.64	bcd
12	8.13	cd
13	8.04	cd
14	7.21	d
15	6.50	d
16	7.86	cd
17	7.00	d

The height of the first fruit-carrying truss was significantly lower in the GHs with white mulch as compared to those with black mulch (Table 34). At $\alpha < 0.1$ level, the percentage of parthenocarpic fruits was significantly reduced in houses with white ground cover.

The time span until the first flowers opened was reduced in houses with black ground mulch by around 2 days (Table 34) compared to the houses with white ground mulch. The roof pigment did not influence the time span.

The time which flowers needed to fully open and to subsequently develop into ripe fruits was neither affected by whether or not the roof plastics were coated with the NIR-reflective pigment paint or by the color of the mulch used to cover the GH floor (Table 34).

Table 34: Average number of inflorescences, height of the first fruit carrying truss, time needed for the first flower to fully open, percentage of parthenocarpic fruits, and ripening time of the first fruits of tomato plants (variety FMTT260) grown in greenhouses with four different set-ups during the dry season 2006 in Central Thailand. Different lower or upper case letters within rows indicate significant differences at $\alpha < 0.05$ or $\alpha < 0.1$ level, respectively (SNK test).

Mulch color		White		White		Black		Black	
Roof is NIR-		transmissive		reflective		transmissive		reflective	
No of inflorescences	n = 41	14.17	a	13.36	a	14.17	a	14.92	a
Height of first fruit-carrying truss [cm]	n = 48	50.00	a	46.75	a	54.42	b	52.25	b
Time needed for the first flower to fully open [days]	n = 48	27.00	a	26.00	a	24.00	b	25.00	b
Proportion of parthenocarpic fruits [%]	n = 111	51.18	A	53.11	A	51.49	B	63.00	B
Time until first ripe fruits occurred [days]	n = 41	37.92	a	38.25	a	38.33	a	40.18	a

The longitudinal growth of the plants grown above black mulch was accelerated from the beginning (Figure 28). Differences in plant height were at significance level $\alpha < 0.1$ in cw1 and cw18 and $\alpha < 0.05$ level in all other weeks. Again, the roof cover did not exert any effect.

In the second half of the experiment with beginning of the hot-dry season (cw12) the effects of the different GH set-ups on plant growth became more pronounced resulting in bigger differences between the houses. The maximal differences between the two houses 'bref' and 'wref' are shown in Figure 29. Stronger effects resulted in a steeper slope of the curve from cw9 onwards, implying that the effectiveness of the white mulch increased with ongoing trial duration (coinciding with increasing global radiation intensity and hence air temperatures).

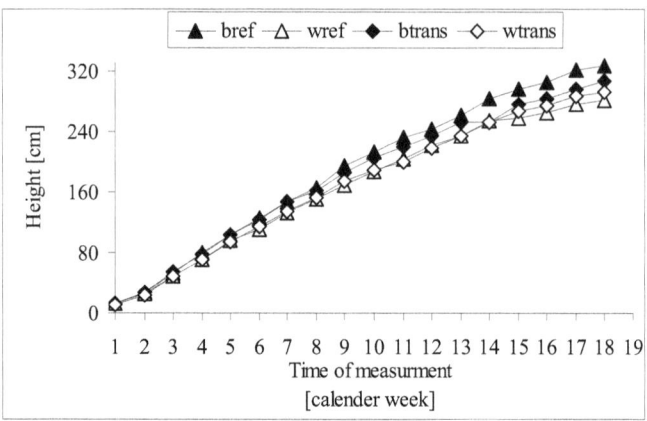

Figure 28: Average plant heights of tomato plants (variety FMTT260) grown in greenhouses (GHs) with different set-ups (bref: black ground mulch/ NIR-reflecting roof cover; wref: white ground mulch/ NIR-reflecting roof cover; btrans: black ground mulch/ NIR-transmissive roof cover; wtrans: white ground mulch/ NIR-transmissive roof cover) during the dry season 2006 in Central Thailand. From calendar week 2 to 17 the differences between GHs with white and black mulches were significant (SNK test, $\alpha < 0.05$, n= 216).

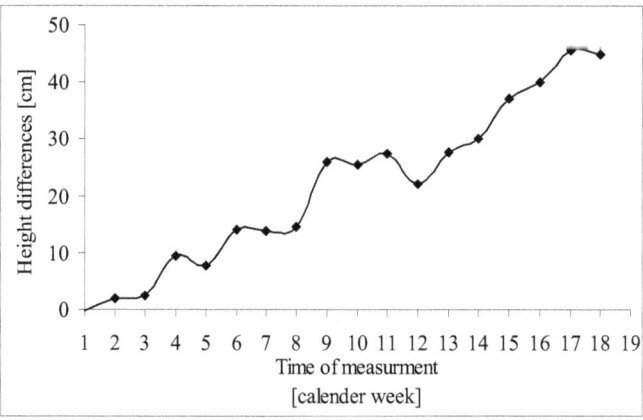

Figure 29: Difference in plant heights of tomato plants (variety FMTT260) grown in greenhouses (GHs) either equipped with black or white mulch (both GHs had NIR-reflecting roof covers) during the dry season 2006 in Central Thailand. The values differed significantly throughout the experimental duration (SNK test, $\alpha < 0.05$, n=108).

Results

No significant differences between the different greenhouses were found regarding the mean numbers of fruits per harvest, total number of harvested fruits, and the number of marketable fruits (Table 35). However, the highest number of fruits was harvested in 'btrans' and ranged from 614 in 'bref' to 711 in 'btrans' (Table 35). The number of non-marketable fruits was higher in the GHs with black mulch than in GHs with white mulch.

The percentage of marketable fruits was significantly higher in the GHs with white mulch compared to those with black ground cover while the roof did not influence this trait. Consequentially, the percentage of parthenocarpic fruits was higher in the GHs with black ground cover than with white mulch.

Table 35: Average number of fruits, sum of fruits, marketable and non-marketable fruits per inflorescence, and percentages of marketable and parthenocarpic fruits of variety FMTT260 in greenhouses with four different set-ups evaluated in the dry season 2006 in Central Thailand. Different lower or upper case letters within rows indicate significant differences at $\alpha< 0.05$ or $\alpha< 0.1$ level, respectively (SNK test, number of fruits, mean number of marketable and non-marketable fruits, percentages of marketable fruits, and percentages of parthenocarpic fruits: n= 778; the sum of fruits: n= 48).

Mulch color	Black		White		Black		White	
Roof is NIR-	reflective		reflective		transmissive		transmissive	
Fruits per inflorescence	3.18	a	3.27	a	3.65	a	3.34	a
Sum of fruits	614.00	a	647.00	a	711.00	a	641.00	a
Marketable fruits per inflorescence	1.64	a	1.86	a	1.91	a	1.95	a
Non-marketable fruits per inflorescence	1.54	a	1.40	b	1.74	a	1.39	b
Marketable fruits [%]	52.27	a	58.28	b	53.55	a	60.75	b
Parthenocarpic fruits [%]	10.77	A	5.10	B	7.66	A	8.00	B

From cw11 onwards, the number of parthenocarpic fruits increased steadily until the end of the experiment (Figure 30). During the last three weeks of the experiment (cw16 to cw18) the proportion of parthenocarpic fruits increased significantly from one week to the next.

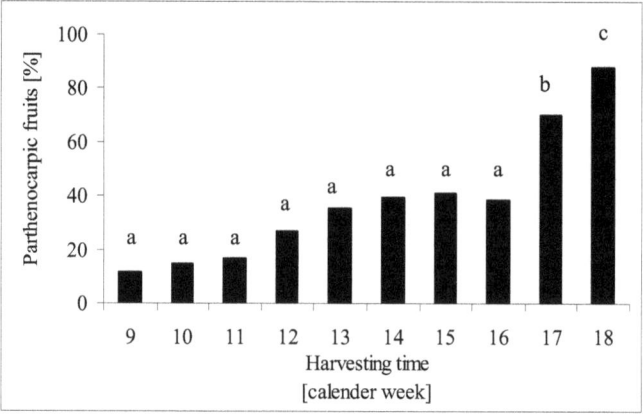

Figure 30: Proportion of parthenocarpic fruits harvested from tomato plants (variety FMTT260) grown in greenhouses during the dry season 2006 in Central Thailand. Values are means across 4 different greenhouse set-ups. Different letters above the columns indicate significant differences between harvesting dates (SNK test, $\alpha < 0.05$, n= 111).

The total yield did not differ significantly in the GHs with different set-ups (Table 36). However, the percentage of marketable yield was higher in GHs with white ground mulch compared to GHs with black ground mulch (Table 36).

Neither the mean weight per fruit of marketable or non-marketable fruits differed significantly (Table 36).

Likewise, the fractions misshaped fruits and fruits infected with BER did not differ significantly (Table 36).

Nevertheless, the percentage of cracked fruits was significantly higher in houses with NIR-reflecting pigments on the roofs (Table 36).

The percentages of marketable yield and non-marketable yield were significantly higher in GHs with white and black white ground mulch, respectively (Table 36).

Table 36: Total yield (Ytot), marketable yield (Ymark), marketable yield per fruit (Ymark fr^{-1}), percentages of the marketable (Frmark) and non-marketable yield (Frnon), misshaped fruits (FrMis), fruits infected with blossom end rot (BER) (FrBER) and cracked fruits (FrCr) of variety FMTT260 grown in greenhouses with four different set-ups evaluated in the dry season 2006 in Thailand. The set-ups differed in their mulch color (black ['b'] or white ['w'] and their roof properties (NIR-transmissive ['trans'] or coated with NIR-shading paint ['ref']). Different letters within a column indicate significant differences between the GH set-ups (SNK test, a< 0.05, total yield/ marketable yield/ marketable yield per fruit n= 912; percentages marketable/ non-marketable yield: n= 778; percentages misshaped fruits/ fruits infected with BER/ cracked: n=506).

Set-up	Ytot [kg]	Ymark [kg]	Ymark fr^{-1} [kg/fruit]	Frmark [%]	Frnon [%]	FrMis [%]	FrBER [%]	FrCr [%]
,bref'	0.22 a	0.12 a	0.05 a	56.20 a	43.87 a	56.44 a	33.14 a	11.25 a
,wref'	0.25 a	0.14 a	0.06 a	61.40 b	38.74 b	55.35 a	21.44 a	23.86 a
,btrans'	0.26 a	0.14 a	0.05 a	57.13 a	43.15 a	51.95 a	30.03 a	17.94 b
,wtrans'	0.24 a	0.15 a	0.06 a	63.99 b	36.07 b	59.36 a	34.15 a	7.16 b

The total yield differed significantly at different times of harvest. The temporal development of the total yield during the course of the experiment was rather ambiguous: yield peaks and depressions followed each other in a biweekly or weekly rhythm in the first 4 harvesting weeks and in the last six weeks, respectively (Figure 31).

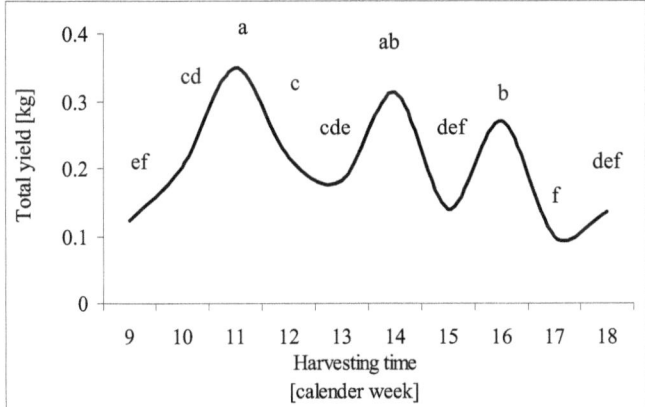

Figure 31: Total yield [kg] harvested from tomato plants (variety FMTT260) grown in greenhouses during the dry season 2006 in Central Thailand. Values are means across 4 different greenhouse set-ups. Different letters above the data points indicate significant differences between harvesting dates (SNK-Test, α < 0.05, n= 912).

Results

For further investigations an experiment was conducted from cw32/2005 to cw38/2005 in order to evaluate the influence of different cooling methods with special emphasis on pollen viability and pollen amount. Therefore, each three plants of the cultivars 2-10 (listed in Table 23) were transplanted in three different GH types. Two of the GHs were clad with 72mesh nets on the sidewalls while one was clad with 50mesh nets. The roof of one of the houses with 72mesh clad sidewalls was additionally coated with the NIR-reflecting pigment. Furthermore, three plants of the hybrid FMTT260 were planted in an additional GH equipped with an evaporative ('fan and pad') cooling system (FAP).

In all greenhouses the floor was covered with the white surface of the PE-mulch upside.

Neither the mesh sizes of the sidewalls nor the NIR-reflecting pigment or the cooling system affected the pollen viability (Table 37) of variety FMTT260. Nevertheless, the highest pollen viability was measured in the house with 72mesh sidewall nets and the NIR-transmissive roof cover.

The highest pollen amount was harvested in the fan-and-pad cooled GH (FAP) and the lowest in the GH with 72mesh sidewall nets and NIR-reflecting roof cover (Table 37).

Table 37: Average proportion of viable pollen (Poll) and pollen amount per flower (Poll Fl^{-1}) (variety FMTT260) grown in four greenhouses with different set-ups in the rainy season in Thailand in 2005. The greenhouses differed in the mesh sizes of the sidewalls (72mesh ['72'] or 50mesh ['50']), the roof properties (coated with NIR-reflecting pigment ['ref'] or NIR-transmissive ['trans'], and the cooling system (natural or fan and pad cooling system [FAP]). Different letters within columns indicate significant differences between the set-ups (SNK-Test, $\alpha < 0.05$, for the pollen viability: n= 13; for the pollen amount: n= 64).

GH set-up	Poll [%]		Poll Fl^{-1}	
'72/ref'	8.73	a	63.31	a
'72/trans'	12.65	a	280.72	ab
'50/trans'	6.10	a	692.40	ab
FAP	0.00	a	4,994.29	b

The genotype (Table 38) as well as the sampling date (Table 39) affected the pollen amount of nine genotypes measured in GHs with different set-ups (50/ 72mesh size sidewalls, NIR-reflecting/ -transmissive roof) in the rainy season 2005. The five lines provided by the AVRDC performed best and showed the highest pollen release. The highest pollen amount per flower was obtained in CL5915 with around 18,747. The lowest pollen amount was found in LA3320 with only one pollen grain.

Table 38: Pollen amount per flower (Poll Fl^{-1}) of nine tomato genotypes grown in the rainy season in Thailand in 2005 averaged across three GHs with different set-ups. Different letters within columns indicate significant differences between the genotypes (SNK-Test, α< 0.05, n= 186).

Genotype	Poll Fl^{-1}	
CL5915	18,747.42	a
CLN2001A	9,860.50	ab
CLN1621L	3,426.63	ab
CLN2418A	359.70	bc
FMTT260	298.24	bc
LA2661	31.46	cd
LA2662	10.33	de
LA3320	0.78	e

Table 39: Pollen amount per flower (Poll Fl^{-1}) of tomato plants grown in the rainy season in Thailand in 2005 averaged across nine genotypes and GHs with three different set-ups measured at different sampling times (calendar weeks [cw]). Different letters within columns indicate significant differences between the sampling time (SNK-Test, α< 0.05, n= 186).

Sampling time [cw]	Poll Fl^{-1}	
32	2,172.52	a
33	63.69	b
34	667.37	ab
35	2,138.28	a
36	36.37	b
37	47.73	b
38	358.64	ab

The different set-ups did not influence the pollen amount. However, the pollen release was highest in GH '72/ref' and lowest in GH '50/trans' (Table 40).

Results

Table 40: Pollen amount per flower (Poll Fl-1) of tomato plants grown in three greenhouses with different set-ups in the rainy season in Thailand in 2005 averaged across nine different genotypes and seven weeks. The greenhouses differed in the mesh sizes of the sidewalls (72mesh ['72'] or 50mesh ['50']) and the roof properties (coated with NIR-reflecting pigment ['ref'] or NIR-transmissive ['trans']) Different letters within columns indicate significant differences between the set-ups (SNK-Test, $\alpha < 0.05$, n= 186).

Set-up	Poll Fl^{-1}	
'72/trans'	238.32	a
'50/trans'	189.31	a
'72/ref'	598.64	a

The pollen amount increased and decreased in a (bi-) weekly rhythm (Figure 32). A sudden break down was found in cw33 (Figure 32). After a significant reduction in cw33 it increased again almost reaching the prior value and dropped again in cw36. In the two following weeks the pollen amount tended to increase though the trend was not significant (Figure 32).

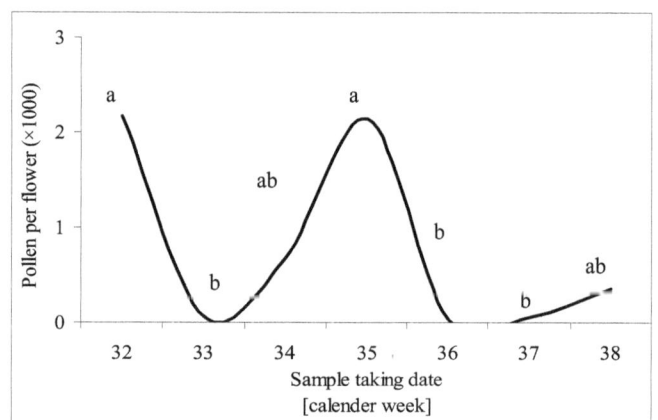

Figure 32: The development of the pollen amount averaged across nine different lines/ varieties and three different greenhouse set-ups (72mesh/ NIR-reflecting roof, 72mesh/ NIR-transmissive roof, 50mesh/ NIR-reflecting roof) grown in Thailand in the rainy season 2005 evaluated in seven weeks (calendar week (cw) 32 to 38) Means with different letters are significantly different between the calendar weeks (SNK-Test, $\alpha < 0.05$, n= 186).

Only five out of nine lines had pollen shed high enough to conduct the staining procedure to assess the pollen viability.

The genotype but not the GH set-ups or the sampling dates affected the pollen viability (Table 41 to Table 44).

Results

The highest proportion of viable pollen was performed in line CLN1621L and was significantly higher compared to FMTT260 (Table 41).

Table 41: Proportion of viable pollen (Poll) of five genotypes evaluated in three greenhouses with different set-ups in the rainy season in Thailand in 2005. Different letters within columns indicate significant differences between the genotypes (SNK-Test, $\alpha < 0.05$, n= 47).

Genotype	Poll [%]	
CLN1621L	42.23	a
CLN2001A	30.98	ab
CL5915	30.77	ab
CLN2418A	26.46	ab
FMTT260	8.97	b

Consistently the percentage of viable pollen tended to be lower in the GH clad with 72mesh sidewalls and NIR-transmissive roof (21 %) than in the other two GHs (Table 42).

Table 42: Proportion of viable pollen (Poll) of tomato plants grown in three greenhouses with different set-ups in the rainy season in Thailand in 2005 averaged across five different genotypes. The greenhouses differed in the mesh sizes of the sidewalls (72mesh ['72'] or 50mesh ['50']) and the roof properties (coated with NIR-reflecting pigment ['ref'] or NIR-transmissive ['trans']) Different letters within columns indicate significant differences between the set-ups (SNK-Test, $\alpha < 0.05$, n= 47).

Set-up	Poll [%]	
'72/trans'	20.65	a
'50/trans'	27.96	a
'72/ref'	27.71	a

Table 43: Proportion of viable pollen (Poll) of tomato plants grown in the rainy season in Thailand in 2005 measured at different sampling times (calendar weeks [cw]) averaged across five genotypes and GHs with three different set-ups. Different letters within columns indicate significant differences between the sampling time (SNK-Test, α< 0.05, n= 47).

Sampling time [cw]	Poll [%]	
32	29.86	a
33	34.96	a
34	23.52	a
35	24.75	a
36	20.98	a
37	29.87	a
38	21.51	a

3.10 Affirmation of heat stress as reason for reduced plant vitality

An experiment was conducted to prove heat stress as the reason for diminished plant vitality. A selection of genotypes grown under greenhouse conditions in Thailand before were chosen and grown under optimum temperatures of 24/ 20 °C (day/ night). Three plants each of the genotypes Hillmar Hellfrucht, CLN2001A, CLN2418A, FMTT260, Pannovy, CL5915, and CLN1621L were transplanted in a GC in Hannover with a plant density of approx. 2.3 plants m^{-2}.

The seeds were sown in cw20/2006 in tray substrate and transplanted in 10 L plastic pots in Potgrond P substrate after two weeks. In cw23/2006, the plants were placed in the GC with controlled temperatures. The indeterminate growing genotypes Hillmar Hellfrucht, CLN2001A, CLN2418A, FMTT260 and Pannovy were pruned weekly while the two remaining varieties with determinate growth habit were not. Plant height was measured once a week and the pollen amount was determined biweekly starting from cw24/2006 until cw31/2006.

Neither the pollen amount nor the weekly height increase differed significantly between the genotypes.

The number of pollen per flower ranged from 61,828,131 in line CLN1621L to 95,266,317 in the hybrid FMTT260 (Table 44).

Table 44: Pollen amount per flower (Poll Fl^{-1}) of five heat tolerant genotypes (supplied by the AVRDC, Taiwan) and the negative control Pannovy (supplied by Syngenta Seeds, Germany) grown under optimum temperatures (24/ 20 °C) in a greenhouse cabinet at Hannover University. Same letters within columns indicate no differences between the genotypes (SNK-Test, α< 0.05, n= 178).

Genotype	Poll Fl^{-1}	
CL5915	130,444.99	a
CLN1621	103,046.88	a
CLN2001	144,941.90	a
CLN2418	147,549.25	a
FMTT260	158,777.20	a
Pannovy	142,031.25	a

The weekly height increase of determinate (Table 45) or indeterminate growing genotypes (Table 46) did not differ significantly. The weekly increment ranged from 19 to 29 cm in indeterminate from 12 to 13 cm in determinate growing genotypes.

Table 45: Weekly height increment (Incr) of three determinate growing heat tolerant genotypes (supplied by the AVRDC, Taiwan) grown under optimum temperatures (24/ 20 °C) in a greenhouse cabinet at Hannover University. Same letters within columns indicate no differences between the genotypes (SNK-Test, α< 0.05, n= 15).

Genotype	Incr [cm]	
CL5915	13.20	a
CLN1621L	12.53	a
CLN2001A	12.00	a

Table 46: Weekly height increment (Incr) of four indeterminate growing genotypes. Two heat tolerant lines (supplied by the AVRDC, Taiwan, labeled by *), the heat sensitive hybrid Pannovy (supplied by Syngenta Seeds, Germany), and the heat sensitive line Hilmar Hellfrucht (supplied by Hild Samen GmbH, Marbach, Germany) grown under optimum temperatures (24/ 20 °C) in a greenhouse cabinet at Hannover University. Same letters within columns indicate no differences between the genotypes (SNK-Test, α< 0.05, n= 20).

Genotype	Incr [cm]	
CLN2418A*	19.27	a
FMTT260*	29.00	a
Pannovy	24.90	a
Hillmar Hellfrucht	24.53	a

No differences of the height of the first developed inflorescences in determinate and indeterminate genotypes were found. Within the determinate lines (Table 47) the height of the first inflorescence ranged from 17 to 26 cm while it was between 32 and 52 cm in indeterminate growing lines (Table 48).

Table 47: Height of the first developed inflorescence of three determinate growing heat tolerant genotypes (supplied by the AVRDC, Taiwan) grown under optimum temperatures (24/ 20 °C) in a greenhouse cabinet at Hannover University. Same letters within columns indicate no differences between the genotypes (SNK-Test, α< 0.05, n= 9).

Genotype	Height [cm]	
CLN1621L	26.00	a
CL5915	25.67	a
CLN2001A	17.33	a

Table 48: Height of the first developed inflorescence of four indeterminate growing genotypes. Two heat tolerant lines (supplied by the AVRDC, Taiwan, labeled by *), the heat sensitive hybrid Pannovy (supplied by Syngenta Seeds, Germany), and the heat sensitive line Hilmar Hellfrucht (supplied by Hild Samen GmbH, Marbach, Germany) grown under optimum temperatures (24/ 20 °C) in a greenhouse cabinet at Hannover University. Same letters within columns indicate no differences between the genotypes (SNK-Test, α< 0.05, n= 12).

Genotype	Height [cm]	
FMTT260*	51.67	a
Pannovy	47.67	a
CLN2418A*	36.67	a
Hillmar Hellfrucht	32.33	a

3.11 Phenotypic evaluation of a segregating F_2 population for mapping QTLs for heat tolerance

Based on the results achieved from the experiments in Thailand the line CLN1621L was chosen as the heat tolerant parent for a cross with the heat sensitive hybrid Pannovy. The emanated F_1 seed was sown and grown under optimum conditions and plants were allowed to self fertilize. Single inflorescences were covered with porous plastic bags (Baumann Saatzuchtbedarf, Waldenburg, Germany). The holes in the plastic film of those bags were small enough to avoid pollen egression but big enough to ensure gas exchange. The size of the plastic bags did not constrain the development of the flowers inside. Immediately when the flowers were sere the bags were removed to alleviate fruit development. When the fruits were overripe they were harvested and cut, and the seed of all fruits of a single plant with as less surrounding colloidal tissue as possible were put in small metal boxes. Inside these boxes the seeds were submerged and kept for two to three days. When the remaining colloidal tissue was detached from the seeds the solution was rinsed with water and the seeds transferred to paper tissues and separated from each other for drying.

The achieved segregating F_2 population was sown in cw24/2006. In cw28/2006 174 F_2 plants plus two plants of the F_1 generation (propagated vegetatively) and both parents, were transplanted into a GC with 8 × 12 × 4 m (Figure 33) with 32/ 28 °C day/ night temperatures, resulting in a plant density of approx. 1.8 plants m^{-2}. The temperature was measured with a data logging system storing mean data every 12 minutes. Insect pests were controlled once a week. Spraying of fungicides was done according to necessity prior to flowering. Fertigation was done manually in different intervals depending on plant age.

Results

Figure 33: A section of the segregating F_2 population in the greenhouse cabinet just after transplanting. Plants were grown on strings for stabilization at 32/ 28 °C day/ night temperatures for 12 weeks.

The pollen parameters were rated twice weekly and the vegetative data number of inflorescences, flowers and fruits per inflorescence weekly starting from cw29/2007.

Pollen of single flowers of all flourishing plants were collected into 5 ml PE tubes (length: 75 mm, diameter: 12 mm). These tubes were closed tightly with a buckler (Sarstedt, Nümbrecht, Germany). The pollen were mixed with 500 μl FDA staining solution. After five minutes incubation time, pollen were evaluated with the aid of a flow cytometer.

Flow cytometers are able to analyze several thousand particles every second and can actively separate and isolate particles having specified properties. Flow cytometry offers high-throughput combined with automated quantification of set parameters instead of producing an image of the cell like a microscope. A flow cytometer consists of five main components. The flow cell- liquid stream carries and aligns the cells so that they pass single file through the light beam for sensing. This light beams usually are mercury or a xenon lamps and high power water-cooled lasers, and low power air-cooled lasers. Another part is the detector and Analogue to Digital Conversion (ADC) system generating fluorescence signals. It comes along with an amplification system and a computer for analysis of the signals. Modern instruments usually have multiple lasers and fluorescence detectors. Increasing the number of lasers and detectors allows for multiple antibody labeling, and can more precisely identify a target population by their phenotype.

Every sample was investigated for 180 seconds. The laser refraction, the wavelength of the emitted light, and the light diffusion was detected. The data generated by the flow-cytometer were plotted in two dimensions to produce a histogram. The computer output showed a curve according to the light emission of the pollen (Figure 34). On the x-axis the logarithm of the fluorescence light intensity was plotted. Therefore, the bright fluorescent - vital- pollen were characterized by a peak on the right side (high light emission) while the non fluorescent - non-viable- pollen were characterized by a peak on the left side (low light emission). The peak height (y-axis) is caused by the count of pollen in relation to the measured light intensity. Thresholds were defined before the experiment started. The thresholds covered a certain area under the resulting curve and the results were expressed as percentages of pollen per defined class. The threshold for dead pollen was set by dint of heat-slain pollen resulting in one single peak on the left side. Setting the threshold for living pollen was done with the aid of 20 pollen samples grown under optimum temperatures characterized by a high percentage of viable pollen resulting in a weak peak on the left side and a strong peak on the right side. The peaks on the left side for dead pollen were reconciled. All pollen leading to readings between these two peaks were in a third class as slightly degraded pollen. The establishment of these thresholds enabled the definition of three classes of pollen viability. To avoid any effects of time the sequence in which the samples were measured was completely randomized.

The applications for flow cytometers are wide and include various practices in medicine and molecular biology- for determining ploidy levels of cells e.g.- but to our knowledge- no high throughput investigations of FDA stained pollen were published to far. Therefore, the present study is the first report of a successful application of this method for a high-throughput analysis of pollen viability.

Subsequent all samples were counted using the Fuchs-Rosenthal-Chamber to evaluate the pollen amount.

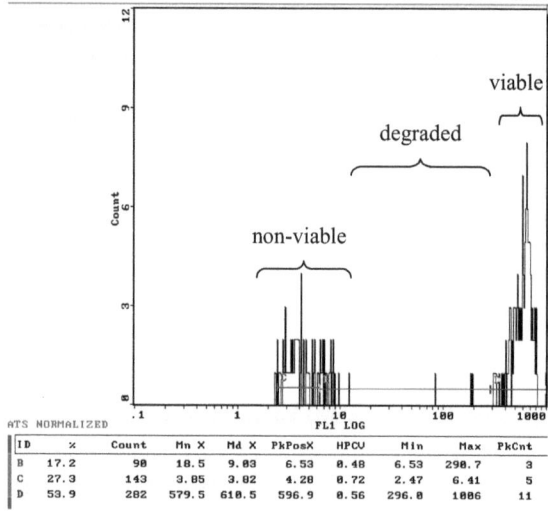

Figure 34: Print out for one pollen sample measured by the flow cytometer. The peaks on the left and right side define the non-viable and viable pollen, respectively.

For all traits evaluated, high variability within the F_2 population was found (e.g. Figure 35). The number of inflorescences, flowers, fruits, fertilized fruits, and the percentages of fruit set and fertilized fruits were significantly different between genotypes of the F_2 population. All values for the 177 genotypes evaluated are listed in the appendix. The number of inflorescences ranged from one to 19. The numbers of flowers and fruits per inflorescence averaged between 6.3 and 10.4 and between 0 and 4.14, respectively.

The percentage of fertilized fruits ranged from 10 % to 100 %.

Genotype 135 produced the lowest number of pollen per flower (2,206) while the reverse was true for genotype 51 (165,000 pollen per flower).

The total number of flowers was correlated with the total number of fruits ($r^2 = 0.46$; $p < 0.0001$). The number of flowers per inflorescence was correlated with their number of fruits per inflorescence ($r^2 = 0.32$; $p < 0.0001$, Table 49). The pollen amount and the percentage of viable pollen did not correlate with the total number of fruits or the mean number of fruits but with the numbers and means of fertilized fruits (Table 49).

The two parental lines did not differ significantly in one of the traits evaluated though the pollen viability of the heat sensitive parent was only half of the viability attained by the heat tolerant parent.

Results

Figure 35: Segregation of the fruit color of unripe and ripe fruits in different genotypes of the segregating population grown under heat stress (32/ 28 °C) in a greenhouse cabinet at Hannover University for 12 weeks. Bright green fruits did not result in bright red fruits necessarily.

Table 49: The Pearson correlation coefficients and significances for the traits flowers (Flav Inf^{-1}) and fruits per inflorescence (Frav Inf^{-1}), sum of flowers (Flsum), fruits (Frsum) and fertilized fruits (Frfert), percentages of fertilized fruits per inflorescences (Frfert Inf^{-1}) and viable pollen (Poll), and the pollen amount per flower (Poll Fl^{-1}). Significances at levels of 0.05 are indicated by ** and not significant results are depicted by n.s.

	Flav Inf^{-1}	Frav Inf^{-1}	Flsum	Frsum	Frfert Inf^{-1} [%]	Frfert	Poll [%]	Poll Fl^{-1}
Flav Inf^{-1}		0.32 **						
Frav Inf^{-1}							0.11 n.s.	0.10 n.s.
Flsum				0.46 **				
Frsum							0.02 n.s.	0.05 n.s.
Frfert Inf^{-1} [%]							0.26 **	0.31 **
Frfert							0.17 **	0.42 **
Poll [%]								0.04 n.s.
Poll Fl^{-1}								

The percentages of viable, slightly degraded, and non-viable pollen ranged from 4 % to 65 %, 15 % to 73 % and 12 to 48 % and varied significantly between different genotypes of the F_2 population. The distributions of the values for viable pollen, the percentage of fertilized fruits, the pollen amount, the numbers of flowers and fruits per inflorescence, and the percentage fruit set are depicted in Figure 36 to Figure 41.

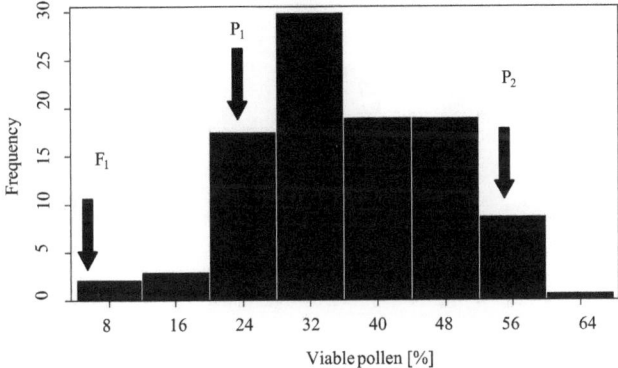

Figure 36: Frequency distribution of the percentage of viable pollen in the segregating population grown under heat stress (32/ 28 °C) in a greenhouse cabinet at Hannover University for twelve weeks. Values for the parents P_1 (heat sensitive), P_2 (heat tolerant), and the first filial generation (F_1) are depicted by arrows.

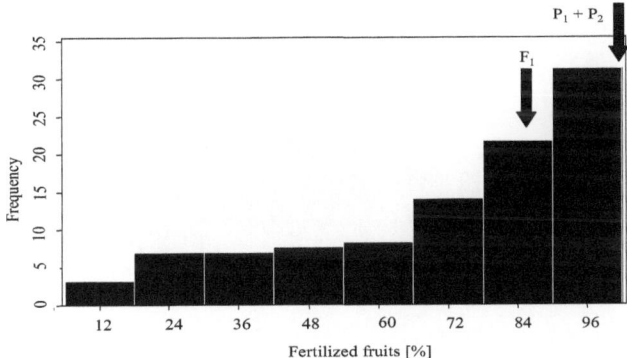

Figure 37: Frequency distribution for the percentage of fertilized fruits in the segregating population grown under heat stress (32/ 28 °C) in a greenhouse cabinet at Hannover University for twelve weeks. Values for the parents P_1 (heat sensitive), P_2 (heat tolerant), and the first filial generation (F_1) are depicted by arrows.

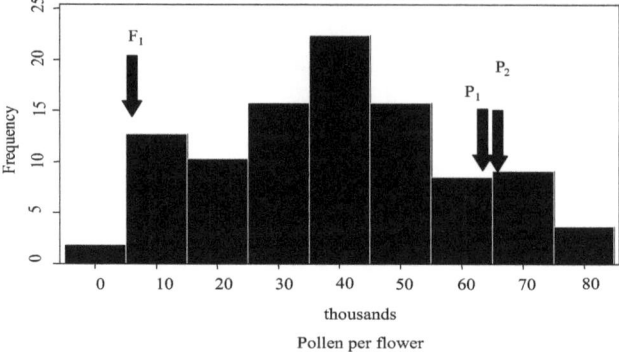

Figure 38: Frequency distribution for the pollen amount in the segregating population grown under heat stress (32/ 28 °C) in a greenhouse cabinet at Hannover University for twelve weeks. Values for the parents P_1 (heat sensitive), P_2 (heat tolerant), and the first filial generation (F_1) are depicted by arrows.

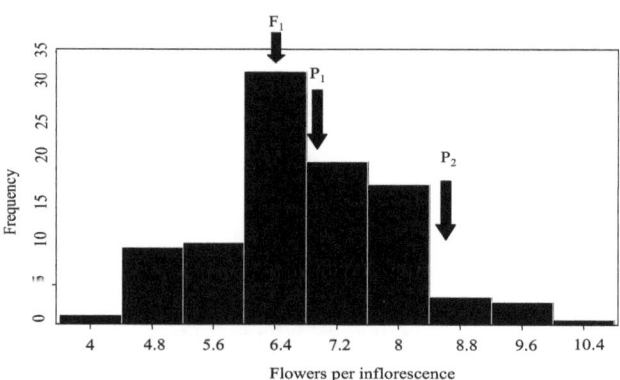

Figure 39: Frequency distribution for the flowers per inflorescence in the segregating population grown under heat stress (32/ 28 °C) in a greenhouse cabinet at Hannover University for twelve weeks. Values for the parents P_1 (heat sensitive), P_2 (heat tolerant), and the first filial generation (F_1) are depicted by arrows.

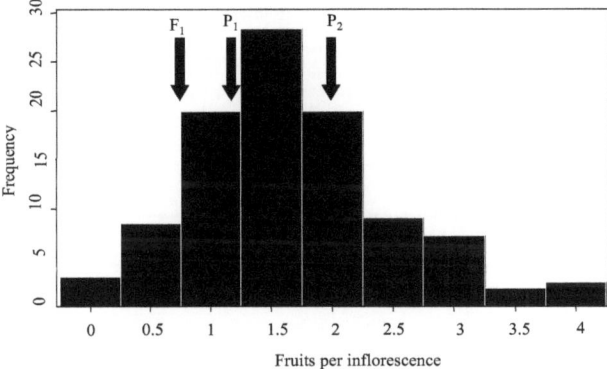

Figure 40: Frequency distribution for the fruits per inflorescence in the segregating population grown under heat stress (32/ 28 °C) in a greenhouse cabinet at Hannover University for twelve weeks. Values for the parents P_1 (heat sensitive), P_2 (heat tolerant), and the first filial generation (F_1) are depicted by arrows.

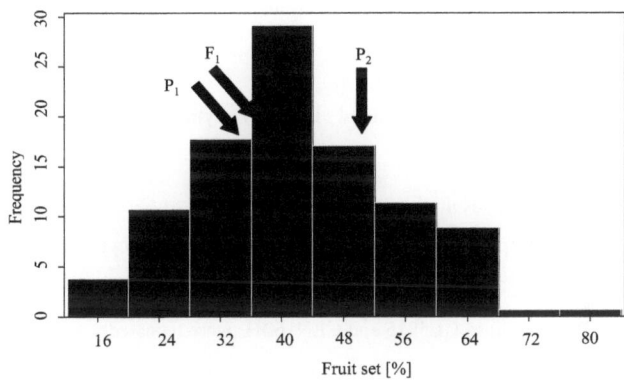

Figure 41: Frequency distribution for the percentage of set fruits in the segregating population grown under heat stress (32/ 28 °C) in a greenhouse cabinet at Hannover University for twelve weeks. Values for the parents P_1 (heat sensitive), P_2 (heat tolerant), and the first filial generation (F_1) are depicted by arrows.

The climate data during the course of the experiment are shown in Figure 42. From the 19.07.2006 onwards the predetermined desired night temperature of 28 °C was consistently one degree lower than the set value.

The night temperature decreased from the 16.07.2006 to the 19.07.2006 to a minimum of 19 °C. At the same time the day temperature dropped to a low mean value of 25 °C approx. 9 °C below the set point. In the ongoing experiment the daytime temperatures reached the predetermined values of 32 °C at the 31.07.2006 and leveled off at around 30 °C subsequently.

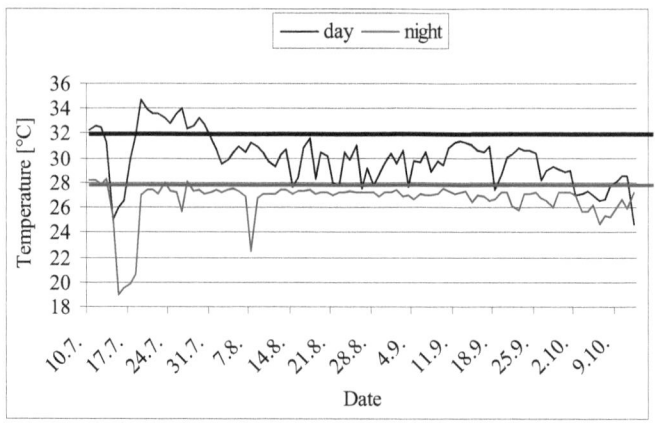

Figure 42: Average temperatures (day and night) in the greenhouse cabinet. The predetermined desired values were 32 °C and 28 °C (depicted by bold horizontally lines).

3.12 Comparison of results obtained by flow cytometry with results achieved by microscopy

For achieving more reliable results basing on higher numbers of observations the manual rating of pollen under the microscope was replaced by the rating via a flow cytometer. The cytometer was able to multiply the number of pollen measured per time unit.

Plants of Pannovy and the lines CL5915 and CLN1621L were repeatedly sown with a time delay of two weeks. Nurturing took place at optimum conditions. After two weeks, the plants were transplanted to GC with 32/ 28 °C or 24/ 20 °C. Pollen from flowers in both GC were harvested directly after anthesis and the samples were divided into two subsamples of almost equal size. One half was stained with FDA and evaluated with the aid of the flow cytometer. The second half was stained with FDA as well but evaluated by microscopy. Both procedures were undertaken as contemporary as possible. Pollen grains of 38 flowers were evaluated for the comparison of the cytometer measuring and the evaluation under the microscope of FDA stained pollen grains.

Since neither the genotypes used nor the temperature influenced the pollen viability the results were pooled for further statistical analyses. For the three class's viable pollen, degraded pollen, and non-viable pollen linear regressions were calculated first. Subsequently the correlations were studied. Results are depicted in Figure 43 to Figure 45.

Results

The Pearson correlation coefficients (r) were 0.75, 0.85, and 0.70 for viable pollen, degraded pollen and non-viable pollen, respectively. The coefficients of determination (r^2) for the respective classes were 0.86, 0.92, and 0.84.

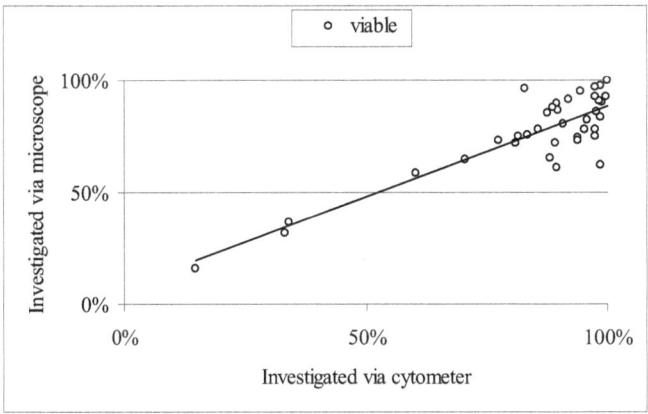

Figure 43: Correlation between results for the pollen viability (by FDA staining) obtained either by microscopy or flow cytometry (r^2= 0.87, p< 0.001, n= 38).

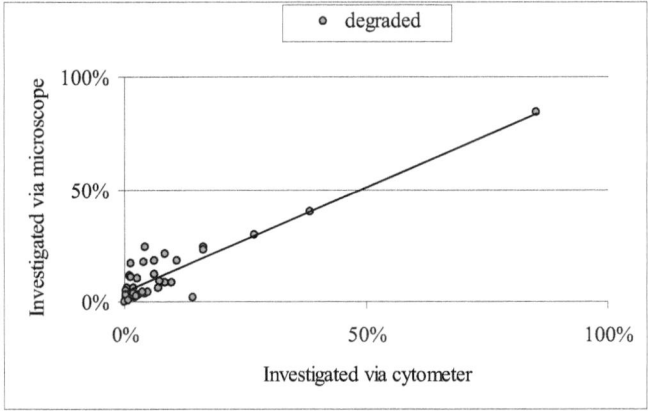

Figure 44: Correlation between results for the pollen viability (by FDA staining) obtained either by microscopy or flow cytometry (r^2= 0.92, p< 0.001, n= 38).

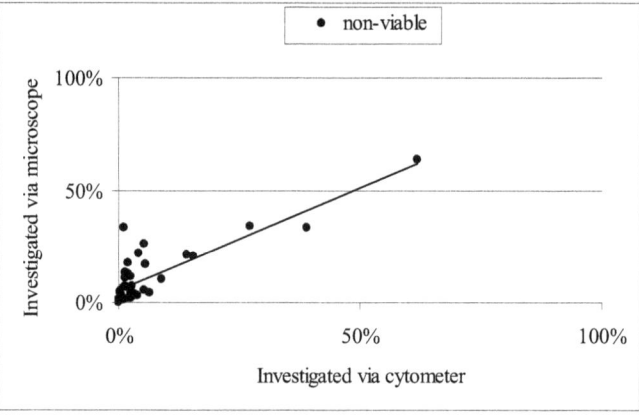

Figure 45: Correlation between results for the pollen viability (by FDA staining) obtained either by microscopy or flow cytometry (r^2= 0.84, p< 0.001, n= 38).

3.13 Analyses of AFLP markers within the segregating F_2 population

To detect molecular markers linked with the evaluated phenotypic data amplified fragment length polymorphism (AFLP) analyses were conducted according to two different protocols.

The AFLP analyses conducted according protocol 1 did not give any reliable results (Figure 46).

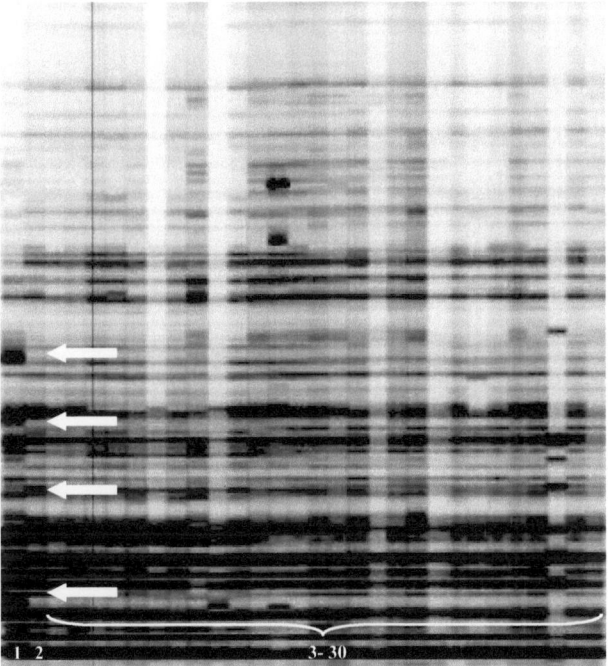

Figure 46: Detail of a band patterns of 29 genotypes after the accomplishment of AFLP PCRs. PCR products were separated on a polyacrylamide gel and visualized on a DNA analyzer. All genotypes were restricted with the *Tru*I/ *Mse*I enzyme combination and PCRs were conducted with the same primer combinations. The lanes one and two contain PCR products of two independent AFLP reactions based on DNA of the same genotype. The lanes three to 30 contain PRC products of different genotypes of the F_2 population. Differences between the repeated genotype were depicted by arrows.

Repeated experiments with the same primer combinations resulted in different band patterns of the same genotype investigated. The enzyme/ primer combinations were found to be unusable for analyses of the plant material used.

The protocol 2 was used to test the reliability of the method in a small sub-fraction of the population so far. With the enzyme/ primer combinations used (see appendix) the band patterns of the same genotypes were identical and some polymorphisms were found Figure 47.

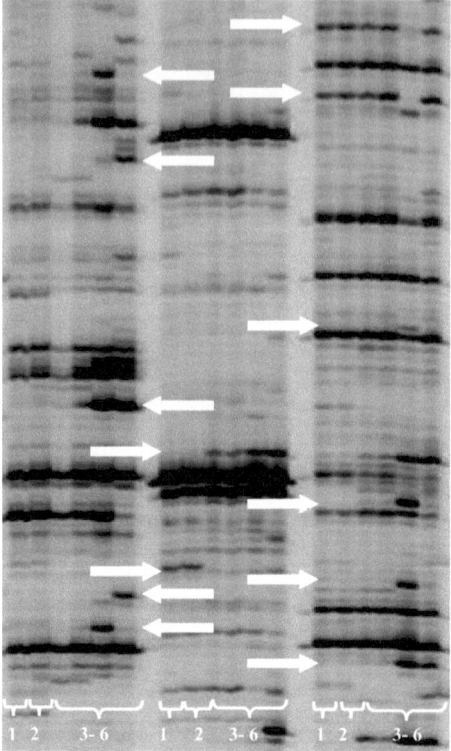

Figure 47: Detail of a band patterns of five genotypes after the accomplishment of AFLP PCRs. PCR products were separated on a polyacrylamide gel and visualized on a DNA analyzer. All genotypes were restricted with the $EcoRI/MseI$ enzyme combination and PCRs were conducted with three primer combinations. The lanes one and two of each of the three compartments display AFLP products based on the DNA of the same genotype while the lanes three to six contain PRC products of different genotypes of the F_2 population. Markers differentiating between genotypes are depicted by arrows.

3.14 Production of F_3 seed for further investigations

To alleviate the disadvantages of the F_2 population and create a F_3 population side branches from every plant of the segregating population were taken to propagate them vegetatively. Since no living plant material of the F_2 can be preserved for ongoing experiments a next generation was produced for the price of less segregation.

At least five cuttings per plant were done. The cutting area was pruinose with growth hormone and they were transferred into substrate. The cuttings were covered with plastic domes to increase RH for one to two weeks. The cuttings were grown under optimum conditions. After the plants started flowering they were selfed to produce F_3 seeds as described before and the seeds stored in plastic boxes with silica gel.

4 Discussion

Heat stress is a problem for agriculture in many areas of the world. With regard to the debate on global warming this topic is also attracting more and more attention in regions with moderate climate.

Many experiments under heat stress have been conducted already. Most of these experiments were carried out under heat shock treatments. Heat shock is usually defined as the short time exposure of plants to temperatures significantly above the upper temperature threshold for plant growth leading to direct injuries of plants including protein denaturation and aggregation or increased fluidity of membrane lipids (Wahid et al., 2007b).

Since around ten years experiments focused more on moderately elevated temperatures (Peet et al., 1997; Sato et al., 2006) leading to indirect injuries. These indirect injuries include the inactivation of enzymes, the inhibition of protein synthesis, protein degradation and losses of membrane integrity (Howarth, 2005).

Reported alterations caused by heat stress associated with the fruit set in tomato are mainly changes and malfunctions of reproductive organs (Charles and Harris, 1972; Dane et. al., 1991; Peet and Bartholemew, 1996). Nevertheless, most researchers found poor fruit set attributed to more than one stressor (Rudich et al., 1977; Kuo et al., 1979). Furthermore, different plant species and plants of one species differed in their response to heat stress at diverse developmental stages (Wahid et al., 2007b).

Intend of this study was therefore to identify factors limiting fruit set in tomato under heat stress in tropical climate regions. Results were considered important for the improvement of heat tolerance in tomato for further breeding purposes and scientific research.

4.1 Methods

4.1.1 Plant material

Partially, the selection of the plant material used in experiments was done *in silico* basing on the plant breeders/ collectors descriptions.

The descriptions of the TGRC's stock list mentioned heat tolerance for all accessions chosen. Therefore, the poor performance of these varieties in the accomplished experiments was striking. In all experiments under heat stress irrespective whether they were conducted in climate chambers in Hannover or in greenhouses in Thailand the lines from TGRC achieved inferior results for all traits investigated compared to the heat tolerant lines from AVRDC. Many times results were even worse compared to the negative control.

A potential reason for this misdemeanor might be the origin of the varieties supplied by the TGRC. Since the varieties were bred and propagated in a hot but dry climate zone they may not be able to scope with the high humidity of the tropics.

Likewise, the heat tolerant hybrid HT7, the first nationally recognized hybrid tomato from Vietnam, yielded poorly for several traits investigated (the number of flowers per inflorescence, the pollen amount, and the percentage of parthenocarpic fruits) compared to heat tolerant lines from AVRDC and the negative control in GHs in Thailand. Since HT7 was developed by the Center for Research and Development of High Quality Vegetable Varieties in Hanoi (FAO database Statistic, 2000) with similar climate compared to Thailand it is unlikely to hold its origin responsible for the inferior development. HT7 was developed for field production systems and probably the selection for traits important for field production caused an aggregation of traits disadvantageous for greenhouse production.

In general, it was difficult to cultivate determinate growing genotypes in GHs or CC. Through their bushy habits the plants needed more space compared to indeterminate growing plants. The additional space was not considered in the planning of the placement of the plants in the GH. Neighboring plants of the same and of different rows grew together and overgrew the supply paths. This led necessarily to premorse branches and potentially in misinterpretations of total inflorescence numbers. However, the loss of some branches of determinate genotypes was neglected for the reason that they built several hundreds of inflorescences and the evaluation was restricted to fifty inflorescences for analyses in case plants surpassed this number.

Furthermore, the height measurement of determinate plants and its comparison to plant heights of indeterminate plants was complicated. Therefore, comparisons especially of the plant height were done between eitherdeterminate or indeterminate genotypes, respectively, or the weekly height increase was used for analysis instead of total plant heights.

For breeding purposes the determinate growth might be disregarded since it is a monogenic inherited recessive trait and can thus be eliminated in offsprings through few crosses.

The choice of the variety Pannovy as heat sensitive parent for the creation of the segregating population was done assuming that no selection for heat tolerance has taken place in a European high performance variety and because of its availability. The disadvantage of the choice of a F_1 hybrid was the occurrence of segregation in the offspring from the cross of Pannovy and the heat tolerant line. The heterosis effect might partially explain the exceptional good performance of

Pannovy under heat stress compared to the heat tolerant lines and the transgressive segregation of the offspring.

4.1.2 Pollen counting

The evaluation of the pollen amount is an estimation of the pollen number produced in the anthers. Since pollen produced but not released are of no relevance for fertilization and they did not correlate with the fruit set (Sato et al., 2006)the focus in our experiments was laid on the pollen release. Potential influences of changing environmental conditions such as air humidity, light spectra or intensity, and the turgor of the anthers on the pollen release were minimized by collecting all samples at once. The time span of five seconds shaking was increased to ten seconds in order to extract as much pollen from the anthers as possible and thereby to reduce the standard error as much as possible. For the same reason the batteries of the toothbrush were checked and charged regularly to ensure similar shaking frequencies.

Though it was tried to keep the sampling procedure as uniform as possible, problems might have occurred due to high plant densities. Though sampling was done before daily nurture procedures were accomplished gardeners or experimenters might have shaken plants while entering the supply paths or greenhouses.

The application of insecticides or fungicides also might influence the pollen formation and/ or release from individual flowers. Fungicide spraying was observed in earlier experiments to exert a negative influence on the development of already opened flowers though it was not investigated systematically. To ensure reliability the application of fungicides was restricted to the period prior to anthesis therefore.

In Hannover as well as in Thailand several plants had to be removed from experiments because of virus infections. It is unknown whether or to what extends virus infections influence pollen development or pollen release but it appears to be possible, that viruses might affect pollen development negatively. However, since symptoms of viral infections become visible only after an incubation period, it appears to be likely that some samples might have been taken from plants already infected with viruses. Data evaluated before the removal were used for statistical analyses when they did not differ from data taken prior to the start of the incubation period.

4.1.3 Pollen staining

In many publications the control of pollen viability was conducted using pollen germination media (Abdul-Baki et al., 1995; Peet et al., 2003; Sato et al., 2006).

Studies with *Brassica napus* suggested that reduced pollen germination rather than pollen viability under high temperature is the major cause of low pollen fertility (Young et al., 2004). Prasad et al.

Diskussion

(2006) and Aloni et al. (2001) demonstrated for peanuts and bell pepper, respectively, high correlations between *in vitro* pollen germination and fruit-set/ seed-set under high-temperature conditions. In the current study, different pollen germinating media were tested. Hedhly et al. (2005) already conveyed of problems correlating *in vitro* pollen germinating with pollen performance *in vivo*. Our results confirmed their statements. No correlation between the pollen germination rates *in vitro* and the pollen viability tested by FDA staining according to Heslop-Harrison et al. (1984) was observed independent of the medium used.

Since pollen determined viable by FDA staining correlated well with the percentages of fertilized fruits the method was supposed to be reasonable.

A problem of the FDA staining procedure was the differentiation between viable pollen (bright fluorescent) and non-viable pollen (non- fluorescent) as done by Heslop-Harrison et al. (1984), Ercan et al. (1996), and Deutsch (2004). In all experiments accomplished, a third class was found whose brightness was intermediate. This class was assumed to be still viable since esterase activity must have led to the elimination of the fluorchrome with subsequent fluorescence. Esterases are usually only functioning in living cells with intact cell walls. Therefore, the fertilization capability of these intermediate fluorescent cells was unclear. Since pollen grains traverse an aging process as all living cells do, the distinction of three classes seemed reasonable though the existence of this third class was – to our knowledge- never described in the literature before.

By correlations of the proportion of each of the three viability classes with the percentages of fertilized fruits the fertilization capability was evidenced. Only bright fluorescent pollen were determined to be able of fertilization.

A technical problem of the FDA staining procedure was the fainting of the fluorescence under UV exposure with time. It was impossible to microscope and to evaluate 100 pollen grains necessary for the achievement of reliable results before fading started. Therefore, photos were taken from the samples exposed to UV light contemporary and the pollen viability was evaluated *in silico*. The brightness of the pollen was rated with the aim of imaging software. The reduction of stray light and adjustment of the background of all pictures provided better comparability. Nevertheless, the determination of the color was still a personal impression perceived. Consequently, only a single person evaluated the samples of one sampling date. Since stray light could not be avoided completely while taking pictures the observed color depended additionally on surrounding light conditions.

It was tried to substitute the UV depending staining method because UV lamps for microscopes are expensive and not available for every microscope. A good alternative was found with the MTT staining procedure confirming earlier results of Khatun et al. (1995) and Rodriguez-Riano et al. (2000).

Regardless the various methods used to evaluate pollen viability the high standard deviation for this trait could not be reduced and the work load was the same with MTT staining compared to FDA staining evaluating the pollen grain color by microscopy.

Therefore, the staining with fluorophores and a subsequent evaluation of the pollen grain color by flow cytometry was implemented. To our knowledge- a high throughput evaluation of pollen viability detecting fluorescence by flow cytometry was never described before. Methods described earlier like the usage of photomultiplier (Heslop-Harrison et al., 1984) or the microscopy of pollen samples (Deutsch et al., 2004) were simplified using the flow cytometer allowing a manifold increase of the number of pollen evaluated per sample in less time. The measurement of the light intensity became independent of personal perception and surrounding light conditions.

4.2 Effects of heat stress on vegetative growth

Optimum temperatures for the net assimilation rate and vegetative growth in tomato were reported to be between 25 and 30 °C (Khavari-Nejad 1980). But diverse optimum temperatures for fruit set were testified by Charles et al. (1972): 18 to 20 °C, de Koning (1994): 21 to 24 °C and Peet et al. (1996): 22 to 25 °C, indicating more complex relationships. Apparently, not only the temperature influences the fruit set but other aspects are of importance.

All authors used different varieties for their heat stress experiments with only some varieties known or supposed to be temperature tolerant. All experiments were conducted under special settings and were not comparable to our purposes and experimental conditions. Since all reports showed lower optimum temperatures for vegetative growth compared to the threshold temperature for vegetative growth, the first step in the current study had to be the designation of a maximum temperature for the plant material used for subsequent heat stress experiments. The temperature should influence the plant development remarkably but not prevent it entirely.

Considering the temperatures of the dry season in Thailand with monthly mean values of 29- 31 °C and daily maxima of 32- 34 °C (www.worldclimate.com, 18.12.2005) the temperatures in the growth chambers were set to 24/ 20 °C and 34/ 30 °C (day/ night) as optimum temperatures and temperatures to induce heat stress reactions, respectively. It was tried to simulate GH conditions in the climate chambers with dim light in the morning and evening simulating dawn and dusk, a light intensity corresponding to a sunny day, and the RH pre-setting on 60 %.

In contrast to Wahid et al. (2007b) who postulated the threshold temperature for a restriction of vegetative growth in tomatoes of 35 °C the temperature of 34 °C in our experiment in the CC exceeded the maximum growth temperature for the varieties FMTT260 and Pannovy without doubt. Plants did not only stopped generative growth but vegetative growth entirely and the plants showed severe physical disorders independent of the variety or heat tolerance level. The disorders varied from reduced plant growth to strongly involute and stiff leaves, the drop of buds and flowers, and premature withering. Since temperatures of 34 °C in GHs did not affect these traits in such a manner other aspects in combination with high temperature or by themselves had to cause these disorders.

Surprisingly we also noticed disorders under optimum temperatures: involute leaves in both FMTT260 and Pannovy and elongated stigmata in Pannovy but not in FMTT260. This indicated other factors than temperature not considered in the experiment but important for plant development as well as genotypic variations in the response to stressors.

Genotypic variation of the response of tomato to heat stress was reported earlier by Wessel-Beaver et al. (1992). Likewise, the problem to uncouple heat stress from other factors as RH or light intensity was demonstrated by Peet et al. (2003). Inside the CC the temperature kept up to the pre-settings but the RH in both chambers never fall below 80 % and most of the time the RH averaged 90 % due to technically constructional defects.

Peet et al. (2003) reported of an increased temperature sensitivity of tomato at a low vapor pressure deficit (VPD). This might explain that we did not find the severe disorders in GHs at similar or higher temperatures but lower RH.

Since the RH in the climate chambers turned out to be uncontrollable the high humidity might have caused the disorders in Pannovy even under optimum temperatures, which were never observed under GH conditions.

In general, high temperature is reported to cause reduced plant height and node lengths (Hall 1990; Ebrahim et al., 1998). In the current study, this was valid for plants of all varieties subjected to heat stress when compared to plants of the same varieties grown under optimum temperature in both, CCs and GHs. Overall, we found less height increment in determinate growing genotypes as compared to indeterminate growing genotypes independent from the temperature.

Since no genotypic variation was found, it yielded the assumption that the reduction of plant growth is caused by general operant mechanisms within plants.

The ratio of red to far red radiation has a significant effect on the partitioning of resources between generative and vegetative plant parts (Crotser et al., 2003). Less height increment in determinate growing plants was often associated with higher numbers of trusses and flowers per inflorescence compared to indeterminate growing genotypes. Therefore, it might be assumed that determinate growing genotypes are able to use differences in the red to far red radiation ratio better and enhance their reproductive growth compared to indeterminate growing genotypes.

The plant growth of several plant species were reported to raise with increasing temperature until a certain limit (Wahid et al., 2007b). After exceeding the limit, the plant growth decreases with further rising of the temperature (Howarth 2005). In the cited studies, the accelerated growth was explained by faster cell expansion under elevated temperatures within a certain temperature range not exceeding the threshold temperature beyond which heat stress commences.

Comparing the height of plants grown in GHs with different set-ups the biggest plant height and an accelerated flourishing were found in GHs with maximum temperatures compared to each other. In both seasons (dry/ rainy) the tallest plants of variety FMTT260 grew in GHs with black ground mulch. In the rainy season the NIR-shading pigment generated an additional effect on plant growth and plants grown in GHs coated with the NIR-shading paint were taller. It might be assumed the threshold temperature for the vegetative growth was not exceeded in both seasons and the shading paint was efficient in adjusting the inside temperature to optimum in the rainy season.

Since the heights of plants grown in Thailand were reduced compared to plant heights measured under optimum temperatures it seemed more likely that the plant growth was more negatively influenced by other aspects not investigated in the experiment and not influenced by the shading paint.

The influence of the shading paint on the plant growth in the rainy season but not in the dry season might refer to the coating of the roof with the NIR- reflecting pigment immediately prior to the start of the experiment. Around six month after the application no influence of the shading paint on the plant growth could be found anymore in the dry season. A declining effectiveness of the shading paint with time was reported by Mutwiwa (2007). Additionally, the lack of uniformity in the distribution of the shading paint on the roof might have effected uneven distribution of light inside the greenhouse reducing the differences in plant heights between the treatments. Therefore, an influence of the shading paint in general on plant growth in the dry season can not be excluded entirely.

In contrast to Mutwiwa (2007) who found reduced plant growth in GHs coated with the shading paint we found taller plants grown in GHs with the NIR-shading pigment on the roof. Mutwiwa (2007) ascribed the reduced plant height to the reduced PAR transmission in these houses. This reduction yielded lower amounts of energy available for assimilation and subsequent reduced plant growth. Probably the reduced temperature had a stronger positive effect than the reduced PAR availability resulting in the height decrement.

Diskussion

The bigger plant heights in GHs covered with the NIR-shading paint compared to heights measured in GHs with NIR-transmissive roof covers supported results of García-Alonso et al. (2006) who found a height increase in pepper in NIR-shaded GHs. Eventually alterations in the light quality caused differences in plant heights between shaded and non-shaded greenhouses.

During the course of the experiment and with subsequent increasing plant height the influence of the combined cooling methods (ground mulch/ roof cover) on plant height became intensified. In the second half of the experiment the NIR-shading probably influenced the meristems directly not depleted by the wide distance of the roof and the plant tips. Results from the experiment conducted in the rainy season might support this idea because the differences caused by the shading paint were measured in a full-grown crop.

However, no closing opinion could be formed about the effectiveness of the cooling methods of the GHs on vegetative plant growth since the results were not consequent. In general, all methods leading to reduced inside temperatures have to be preferred since high temperatures reduce height increase which is correlated with the number of developed inflorescences. High numbers of inflorescences on their hand are important for sufficient yield.

A probable explanation of differences between the plant development in CCs and GHs might be the extraordinary high RH in the climate chambers. High RH and a consequent low VPD lead to reduced transpiration and xylem transport thus. Consequently, physiological calcium (Ca^{2+}) deficiency occurred in all strong transpiring parts of the plants, e.g. leaves. The low Ca^{2+} uptake and transport associated with the low VPD caused the decrement of the export of carbon from tomato leaves (Dinar et al., 1983). According to Tognetti et al. (1998) plants developed mechanisms for dealing with the accumulation of carbon. Plants are able to store the carbon in forms of starch and sugars. This starch accumulation in the leaves caused involute leaf blades.

Under optimum temperatures, plants were able to compensate the low VPD partly. Most likely the high light intensity equivalent to mean sunlight on an average day counteracted and increased the transpiration again what might explain the minor contortion of the leaf blades grown under optimum temperatures compared to leaves under heat stress. Though ADP-glucose-pyrophosphorylase (AGPase) in leaves needs a light-dependent signal for its activation to channel sucrose toward synthesis of storage starch in leaves it was shown by Kolbe et al. (2005) that sugars provide a second input leading to AGPase redox activation and increased starch synthesis and that they can act as a signal which is independent from light.

Under elevated temperatures, not the low VPD itself but combined with high temperature influenced the Ca^{2+} state within the plant leading to even more erected, involute and stiff leaves observed in all heat stress experiments.

The vigor of plants was less reduced in GHs compared to CCs. The application of sidewall nets with small mesh sizes in combination with white mulch color and the NIR-reflecting pigment was expected to reduce the temperature and subsequent Ca^{2+} deficiencies. Reduced mesh sizes for the sidewalls of the greenhouses resulted in higher air exchange rates and lower temperatures (Harmanto et al., 2006) and RH inside the GHs resulting in higher plant transpiration. In contradiction, sidewalls with higher mesh sizes and lead to reduced insect pest infestations.

Overall, the adjustment of single parameters did not significantly improve plant performances attributed to higher fruit set.

4.3 Effects of heat stress on generative growth

In several studies it was demonstrated that the reproductive development of plants is more sensitive to heat stress than the vegetative growth (Kinet et al., 1997; Sato et al., 2000). The results of the current study are supporting this conclusion. In heat sensitive as well as heat tolerant cultivars, high temperatures negatively influenced all investigated traits related to reproduction. We demonstrated in experiments in CC or under GH conditions that temperatures above optimum reduced the number of inflorescences, flowers, fruits and the fertilization. Every single trait might cause reduced fruit set under heat stress.

4.3.1 Numbers of inflorescences

In contrast to results of Mutwiwa (2007), the current studies did not reveal influences of the GH set-ups on the number of inflorescences. Genotypic differences indicated high heritability for this trait. Since Mutwiwa (2007) found a correlation between the plant height and the number of inflorescences vigorous plant growth under heat stress should be a breeding purpose.

Since plants grown in houses with white mulch colors showed decelerated growth with no reduction in their percentage of marketable yield this might provide the possibility of culture time expansion and less work load due to less plant culture procedures.

4.3.2 Flowers

The reduced flower number found under heat stress supported previous findings of Sato et al. (2006) but are contradictory to results of Peet et al. (1997) who did not find a reduction of the flower numbers.

Though Pannovy was able to develop more flowers compared to FMTT260 in CCs under both, optimum temperature and heat stress, the reduction of the inflorescence numbers, flowers and fruits per inflorescences, and the fruit set was higher. An explanation for the higher number of inflorescences, flowers and fruits might be a stronger heterosis effect in the high performance variety Pannovy. Under GH conditions Pannovy built less flowers and fruits per inflorescence compared to FMTT260. Both results indicate less adaptation of Pannovy to high temperatures combined with high RH than FMTT260 and supported our decision to define Pannovy as heat sensitive.

Since temperatures of 34/ 30 °C in the CC redounded to premature withering and severe flower drop, supporting results of Peet et al. (1997) and Lohar et al. (1998), the reduction of the temperature to 32/ 28 °C effected ceased flower dropping in all subsequent experiments.

Nevertheless, heat stress reduced to 32/ 28 °C still caused severe flower malfunctions and deformities as well as growth reductions. Therefore, the decision to conduct heat stress experiments under controlled condition at this temperature range seemed to be reasonable.

One flower malfunction observed frequently in our experiments was stigma exsertion. The proximity of the stigma and the apex of the anther cone is vitally important for self pollinating plants. The final stigma level depends on the relation of the length of the pistil and the stamen. Rick et al. (1969) and Atanassova (1979) recorded that varieties tending to insertions or exsertions are characterized by differences in their styli length. Fernandez-Munoz et al. (1991) reported of differences in the pistil and stamen development under heat stress at different flower stages. The differences varied from the start of flower opening until the flowers opened completely. Fernandez-Munoz et al. (1991) reported of decelerated growth of the stamens leading to a positive stigma level under heat stress but did not find an influence of the light intensity. Fernandez-Munoz et al. (1991) and Dominguez et al. (2005) agreed with earlier results of Sawhney (1983) that high temperature initiate shortened stamen. Atanassova (1979) argued that in buds enzymes or growth substances are activated by heat stress stimulating the fast growth of the style. In contrast, Burk (1929) and Ruttencutter et al. (1975) found exserted stigmata in flowers developed under low light intensity.

In many genotypes used in the current studies the stigmata protruded the answers under heat stress. Since stigma exsertion was observed in Pannovy under both, optimum temperatures and heat stress, the stigma exsertion seemed to be additionally influenced by other stressors than temperature.

No differences of the stigma levels of FMTT260 were found indicating genotypic variation as mentioned by Fernandez-Munoz et al. (1991). FMTT260 is a variety bred in Taiwan under tropical climate conditions. It should be better adapted to high humidity therefore.

Furthermore, low light intensity has to be excluded as reason for stigma exsertion in Pannovy at optimum temperatures since light intensity was high in the CC.

In contrast to the elongated stigmata of Pannovy in the CCs we found stigma exsertion under high temperature GC conditions caused by shortage of the stamen supporting results of Sawhney (1983) and Fernandez-Munoz et al. (1991). The reduction of the stamen length could not be caused by high

RH. Though RH was not measured continue sly the plants grew in naturally ventilated GC. Regular monitoring displayed RH ranges of 60 to 70 % under which no stigma exsertion was observed under GH conditions in Thailand.

It has to be proposed that neither the temperature nor the light intensity were able to cause style elongation by themselves.

Though the stigma exsertion was not investigated particularly no reduced fruit set associated with exserted stigmata was observed under optimum or elevated temperatures.

4.3.3 Gynoecia

While negative effects of high temperatures on the androecia of flowers are well investigated and evidenced the effects of high temperatures on the gynoecia were uncertain. Processes reported to be adversely affected by high temperature are several stages of the flowers development including the meiosis of pollen and ovule mother cells, the number of pollen grains released, and the pollen germination (Kinet et al., 1997).

Peet et al. (1997) found a significant reduction of tomato fruit set with rising temperatures caused by disorders in ovule development and disturbances of post-pollen production processes while Levy et al. (1978) reported of negligible effects of heat stress on macrospores compared to microspores. Fernandez-Munoz et al. (1991) did not find ovule response to low temperatures indicating a broad temperature tolerance of gynoecia.

In our experiments, however, we excluded the interruption of post-pollen processes and the deterioration of the gynoecia. Gynoecia grown under heat stress and pollinated with living pollen were able to develop into fertilized fruits just like gynoecia grown under optimum temperatures. The genotypes used were of different origin and heat tolerance levels (FMTT260, Pannovy, and CL5915) and the results might be generally accepted therefore.

The integrity of the gynoecia was additionally proved by histological investigations. With none staining procedure conducted modifications of the gynoecia tissues developed under heat stress could be observed independent of the developmental stage of the flower at sampling time.

4.3.4 Androecia

Furthermore, no differences in the anther tissues and the pollen development under heat stress could be found with histological investigations until emergence of mature pollen grains in the loculi. Therefore, no restriction of the pollen release necessary for fertilization due to unopened stomata as proposed by Sato et al. (2002) could be supposed and other factors have to cause the reduced fruit set found in our experiments.

Since a damage of the working ability of the styli and ovaries was excluded as reason for reduced fruit set under high temperature, a lower adhesiveness of the stigmata might have been a cause. Pollen might not be able to stick and germinate on dry stigmata. Early work on wet stigmata showed that the main requirement for pollen germination is the presence of a stigmatic secretion. This role of the stigmatic surface has also been demonstrated by genetic ablation experiments in tobacco, where the impaired pollen grain germination and pollen tube penetration into the transmitting tissue of ablated stigmas was reversed by the application of stigmatic exudates (Goldman et al., 1994).

Several failures - known so far - can happen throughout the stages regulating pollen tube access to the ovule. The process of fertilization may be blocked at the initial steps of pollen hydration and germination on the stigma (Franklintong et al., 1992; Nasrallah et al., 1994; Dickinson, 1995). Other processes disturbed might be the pollen tube growth in the style (Matton et al., 1994) or down into the ovaries (Sage, 1994).

Lower carbohydrate reserves in the pollen grains might lead to faster pollen tube attrition in the styli (Herrero et al., 1996) without being recognized by FDA staining or histological methods used.

A lack of Ca^{2+} either caused by low transpiration due to high humidity or by the closure of the stomata due to long-term heat stress (Banon et al., 2004) also affects fertilization. Ca^{2+} cannot be mobilized from older tissues and redistributed via the phloem. Investigations of gametophytes revealed the need of Ca^{2+} for pollen tube growth (Fernandez-Munoz et al., 1991; Holdaway-Clarke et al., 1997; Schiott et al., 2004) and White et al. (2003) postulated the need of Ca^{2+} for pollen tubes to enter the ovules. A lack of Ca^{2+} in the gynoecium might therefore lead to pollen tubes ending in the ovaries without entering the ovules.

The earlier discussed Ca^{2+} deficiency due to low VPD might therefore not only explain disorders of vegetative plant growth but might be important for generative processes. A lack of Ca^{2+} might result in pollen and ovaries appearing viable but not leading to fertilization under Thailand GH conditions resulting in missing correlations of pollen viability and parthenocarpic fruits.

4.3.5 Fruits

In the dry season, the total numbers of fruits and fruits per inflorescence were not influenced by different GH set-ups. The observed variation therefore was caused by genotypic differences. However, the number and percentage of parthenocarpic fruits were increased in GHs with black ground mulch and consequent higher temperatures compared to GHs with white mulch. The increase of parthenocarpy are in line with Kuo et al. (1989) who reported of increasing parthenocarpy with increasing temperature.

Since damages of the gynoecia and influences of the GH set-ups were ruled out damages of pollen were highly probable.

Probable issues might be the impaired development of pollen leading to insufficient numbers of pollen for fertilization or the quality of developed pollen.

4.3.6 Effects on pollen numbers

The pollen tube is the organ delivering the male gametes to the ovaries of flowering plants and is essential for fertilization therefore. Hormaza et al. (1996) observed a reduction of the pollen tube number growing in the style by the same proportion of the reduction of the pollen load on the stigma surface.

That leads to the conclusion that bigger pollen amounts increase the number of fertilized fruits. Though our histological investigations did not allow a quantification of pollen tubes grown in the styli the examinations left the same impression.

Comparisons of pollen amounts measured in heat stress experiments revealed severe reductions of the pollen release compared to measurements under optimum temperatures.

Sato et al. (2000) described differences in the pollen release between varieties under optimum temperatures while the genotypes grown under optimum temperatures in our studies did not differ in their pollen release. The seven genotypes investigated in GHs and CCs were of different origins and heat sensitivity indicating a certain generality of the results.

The control (optimum) temperatures in the experiments were different. While Sato et al. (2000) chose control temperatures of 28/ 22 °C temperatures used in the current studies were 24/ 22 °C. This might explain discrepancies of the results.

However, under heat stress the reduced pollen release showed strong genotypic variation between heat tolerant genotypes and compared to the negative control. In general, with time of heat exposure and increased plant age the pollen amount was reduced. These results were in agreement with results of Sato et al. (2002).

At 34/ 30 °C no appreciable pollen shed and almost no living pollen grains were found independent of the varieties grown supporting the decision of the temperature reduction for subsequent experiments.

Different mulch colors and sidewall mesh sizes, and the NIR-shading paint were not able to increase the pollen release though a tendency for higher pollen release was found in GHs with black ground mulches and NIR-transmissive roofs. The consequently higher temperatures in these GHs seemed to advantage the pollen release. The pollen release is turgor dependent and since plants were fertigated well an increased transpiration at higher temperatures might have induced higher turgor within anthers and consequent easier pollen release.

In our studies - and to our knowledge for the first time -, oscillating pollen numbers during the experimental time were observed. Though the pollen numbers did never correlate with the total fruit numbers or fruit numbers per inflorescence they correlated with the percentage of fertilized fruits under GH conditions. Sato et al. (2002) described the oscillating of percentages of fertilized fruits though the cycles in their experiments were longer compared to the cycles found for the pollen amount in our studies. Basing on the assumption that reduced pollen amount results in reduced fertilization more severe stress might induce advanced damage and recovery of the pollen amount resulting in a more distinct oscillation.

The pollen release was unaffected by the NIR-reflecting pigment, mulch colors, and sidewall mesh sizes indicating other or additional factors than the temperature and air humidity originating reduced pollen shed.

In the FAP cooled GH the pollen amount was highest compared to all other GH set-ups. It might refer to an advantaged development of pollen with decreased temperature since temperatures in active cooled GHs were considerably lower compared to naturally cooled GH (Mutwiwa, 2007).

This advantaged pollen development happened at the expense of reduced pollen viability.

Though most pollen were shed in the FAP cooled GH none of them were viable.

4.3.7 Effects on pollen viability

At temperatures of 34/ 30 °C a complete loss of pollen viability was observed in CCs. Under reduced heat stress at temperatures of 32/ 28 °C the pollen viability was reduced significantly compared to optimum temperatures. Since a reduction of only 2 °C effected a remarkably increase of pollen viability it might indicate a threshold temperature for pollen survive. This was true for plants grown in CCs but no correlation of the pollen viability to the temperature development in the GH in Thailand could be found in the rainy season.

Like the pollen amount, the pollen viability was reduced with increasing duration of heat exposure.

In the rainy season, the pollen viability was higher in GHs with white mulch and consequent lower temperatures while the reverse was true for GHs with black ground mulch. The NIR-reflecting pigment did not influence the pollen viability. In the dry season the contrary applied: the color of the ground mulch did not generate an effect on the pollen viability while pollen viability was reduced in houses coated with the NIR-shading paint. The positive influence of the white mulch color might be referred to the reduced temperatures due to the higher irradiation reflection of white ground mulch compared to black ground mulch. The lacking influence of the shading paint can not be attributed to heavy precipitation and the washing off of the shading paint of the roofs.

More reasonable seems the explanation of the lacking influence of the shading paint in the rainy season by the weather conditions during the experiment. The precipitation in the rainy season 2005 was exceptional low compared to the long time average (www.tutiempo.net, 27.03.2008). With missing precipitation and cloudiness the solar radiation is higher in the rainy season compared to the dry season since the sun is standing closer to the zenith. Probably the high solar radiation exceeded the capability of the shading paint to cool down the inside temperatures to a level advantageous for pollen viability.

A possible reason for the increased reduction of pollen viability in the dry season might be the increase of UV radiation in GHs coated with the NIR-shading paint as reported by Mutwiwa (2007). UV is known to damage cells due to stimulation of the production of active oxygen species (AOS) that cause oxidative damage (Rijstenbil, 2005). Since the shading paint was applied around six month earlier than the experiment started an enduring damage of the UV-absorbing plastic film has to be supposed.

Since pollen viability of several genotypes did not react on different GH set-ups mistakes in the sampling have to be considered. The flower position in the plant canopy affected pollen viability (Burke, 2001). He demonstrated that pollen harvested from flowers within the canopy in the afternoon possessed normal pollen viability while pollen harvested from flowers at the top of the canopy showed a drastic reduction in pollen viability. This differential response of the pollen may be related to lower temperatures in the microenvironment compared to pollen exposed to entering radiation.

In the dry season, the development of the pollen viability reflected the development of the temperatures well. Assuming the heat sensitive stage of pollen around one week before anthesis (Urban, 2006) the observed drop in pollen viability correlated well with the exceeding of the maximum daytime temperatures of 35 °C one week earlier.

Until temperatures did not exceed 35 °C pollen viability increased or decreased, respectively, with increasing or decreasing temperatures. When ambient temperatures did not fall below 35 °C the pollen viability dropped to a low level and did not recover anymore. These results supported the conclusion of a threshold temperature for plant development of 35 °C mentioned by Wahid et al. (2007b). No consensus was found yet whether maximum temperatures, daily mean temperatures or the relationship of day and night temperatures are responsible for the stop of physiological processes (Peet et al., 1996; Peet et al., 1997; Peet et al., 1998b). It seemed physiological plant states were sensitive to elevated temperatures in a different way. This differentiation did not only vary from species to species but among them (Wahid et al., 2007b). For the pollen viability of tomato a maximum temperature of 35 degrees might be assumed.

In the rainy season a complete loss of pollen viability was observed in all genotypes investigated at the same date. Exactly one day after exceeding substrate temperatures of 32 °C no viable pollen could be harvested anymore. The air temperature averaged out 31 °C as it has done around eight days earlier already without consequent loss of pollen viability. Furthermore, the RH reached its highest value during the experiment so far but sustained at the same level while pollen viability started to recover. Therefore, these climate factors were not reasonable for the severe drop of pollen viability. Since no advanced experiments were conducted concerning the substrate temperatures emphasis should be placed on them further on.

4.4 Genetic variability

At present, it is commonly accepted that the tomato was domesticated from *Solanum lycopersicum* var. *cerasiforme* in the area of the states of Puebla and Veracruz in Mexico (Jenkings, 1948; Rick et al., 1975). The necessity to optimize the yield is stressed, and the cultivation of less productive and disease-susceptible traditional varieties is reduced. The process of the loss of genetic diversity between and within populations of the same species is widely known as genetic erosion and a common problem in tomato breeding (Breto et al., 1993; Cebolla-Cornejo et al., 2007). Therefore, it was surprising to find considerable genotypic variation under heat stress within known or supposed to be heat tolerant *Solanum lycopersicum* lines. Tomato cultivars are believed to be derived from a narrow genetic background and are the result of the cumulative effort of numerous breeders over many years

Usually, genetic variation in nature is liable of continuous phenotypic ranges rather than discrete classes. This genetic variation underlies quantitative traits resulting from the segregation of numerous interacting quantitative trait loci (QTLs), whose expression is modified by the environment.

Since in primary segregating generations like F_2 populations the predominant cause of linkage disequilibrium is the physical linkage of loci, they were widely used for classical linkage mapping. The ability to map and characterize polygenes using marker loci is at its highest in populations derived from controlled matings that have the highest linkage disequilibrium (Tanksley, 1993).

Taking consideration of the genetic variability found in the heat tolerant tomato lines we produced a segregating F_2 population by an intraspecific cross of two *S. lycopersicum* genotypes for the evaluation of various phenotypic traits related to fruit set. F_2 populations are the easiest and fastest to construct mapping populations (Asins, 2002; Collard et al., 2005). They are appropriate for

Diskussion

detecting QTLs with recessive alleles and make it possible to estimate dominance and additive effects (Carbonell et al., 1993) and the degree of dominance (Asins, 2002).

A disadvantage is the impossibility of replications under different conditions and the difficulty of studies of epistatic effects (Asins, 2002). Since no variety showed an absolute superior performance for all traits investigated, the line with best performance for most traits investigated was used as the heat tolerant parent. A F_2 population contains the highest genetic variation in case the parents are homozygote (Tanksley, 1993). In the current study, a F_1 hybrid was used as heat sensitive parent. This had reduced genetic variation in the segregating population and might have caused superior results of the heat sensitive parent due to heterosis.

However, all traits evaluated showed high and continuous variability within the segregating F_2 population and evidenced their quantitative characters.

Contrary to our own preliminary results the parents of the population (heat sensitive and heat tolerant, P_1 and P_2) grown together with its offspring (first and second filial generation, F_1 and F_2) under heat stress did not differ significantly for the pollen viability, pollen amount, flower and fruit numbers, percentages of fertilized fruits, or parthenocarpy.

These conflicting results might refer to the temperature residing below the set threshold of 32/ 28 °C for the whole experimental duration. No possibility was found to increase the inside temperature of the GC. A second highly probable explanation might be that only a single P_1 and P_2 plant each were sampled since the repetitions had to be removed due to virus infections.

Except for the percentages of fertilized fruits the distributions were normally distributed evidencing that many genes were involved in phenotypical trait characterization. However, results of the P_1 and P_2 did not range in the left and right tail of the curve except for the pollen viability. The fruit set and the flowers per inflorescence of the P_1 and P_2 ranged around the mean value and little above, respectively. The F_1 performed worse compared to its parental lines for all traits. Compared to its offspring the values for the P_1 are higher than expected. Since the heat sensitive parent itself is a F_1 hybrid it is supposed to perform better than the mean of its parents due to heterosis.

It might be explained by transgressive variation (Tanksley, 1993)due to unmasking recessive genes or accumulation of complementary alleles at multiple loci inherited from the two parents with negative influence on the traits. Transgressive variation is described for all segregating populations. Since the hybrid parent was highly heterozygote transgression might have occurred in the F_1 already. Another reason might be the continuous vegetative propagation and the associated plant impairment of the F_1.

An aberration from the distribution to normal distribution was found for the fruit fertilization. Both, P_1 and P_2 reached 100 % fertilized fruits though the F_1 and several F_2 plants reached less. This might be excited by the accumulation of negative alleles of the parental lines combined in the offspring.

Since a connection of pollen viability to genes on chromosomes two and five might be assumed according to the results of the investigations of the introgression lines (ILs) special emphasis should be placed on further analysis of this trait.

Despite inferior phenotypes, wild species contain genes that can substantially increase tomato fruit quality. Less surprising was therefore the variation within the introgression lines.

The tomato ILs used in the current study were a sub-set of nearly isogenic lines (NILs) developed by Eshed et al. (1992) through a succession of backcrosses, where each line carries a single genetically defined chromosomal segment from the wild type genome. The ILs, representing a whole-genome coverage of *S. pennellii* in overlapping segments in the genetic background of *S. lycopersicum* cv. M82 were phenotyped first in 1993. Presently the library consists of 76 genotypes (Eshed et al., 1994).

Generally, when using ILs for gene mapping the ILs and the parental lines are grown together and traits are evaluated for both, parents and ILs. When the ILs exceed the results measured in the *S. lycopersicum* parent one can assume that the inserted piece of wild species chromosome contains a gene with positive influence. Since we did not get seed of the parental lines, only comparisons between the ILs were possible.

For most traits evaluated in the present study, at least two lines performed significantly better or worse compared to the other lines indicating a gene or QTL with a huge effects of the traits. In most cases, the insertions of the wild species genome were related to different chromosomes. The number of flowers per inflorescence under heat stress was significantly higher for the lines IL2-3 and IL3-3 with insertions on chromosomes 2 and 3, respectively. Therefore, it might be assumed genes in these chromosomal areas influenced the flower development under heat stress positively. Lines with insertions on chromosomes 1, 3, and 12 achieved the best fruit set. Since the two lines with insertions on chromosome 3 have an overlapping segment, it might indicate a QTL in this chromosomal region.

The percentages of parthenocarpic fruits were remarkable. 32 out of 37 lines produced around 100 % parthenocarpic fruits. Only 2 lines showed significantly better results of 20 and 25 % parthenocarpic fruits. This highly significant reduction of parthenocarpic fruits was attributed to insertions on chromosomes 2 and 5. Though the reduction of parthenocarpy corresponded to low numbers of fruits the fruit set of these lines ranged near the mean value measured in the lines. Since the total fruit number was demonstrated to possess high heritability backcrosses with high yielding varieties should solve the problem and increase the fruit number. This indication of QTLs with strong effects on the percentage of pathenocarpic fruits on chromosomes 2 and 5 has to be considered for further investigations.

Since results for the parents are lacking and no pollen traits have been investigated yet the experiment should be repeated and extended to measurements of further traits, especially pollen traits. Only in comparison with the parental lines and under different growth conditions precise conclusions of the number of QTLs that affect a trait, their mode of inheritance, and linkage relationships are possible.

Molecular investigations of the segregating F_2 population, PCR-based marker analyses, were conducted to find markers linked with the phenotypic traits evaluated. Most molecular maps of tomato are based on interspecific crosses (Paterson et al., 1991; Tanksley et al., 1992; Fulton et al., 1997). Since saturated linkage maps are essential tools for genetic studies like quantitative trait mapping, marker-assisted selection, and positional gene cloning AFLP analyses were chosen because they are considered to be a powerful, reliable and rapid assay (Suliman-Pollatschek et al., 2002).

The AFLP technique, developed by Vos et al. (1995), is based on the detection of genomic restriction fragments by PCR amplification and was applied to various plant species including tomato (Saliba-Colombani et al., 2000; Balatero et al., 2002; Spooner et al., 2005).

In the present study, AFLP analyses did not give reliable results yet as many problems occurred during the experiments. The amount of DNA to extract from the dried leaves was very low in some cases and no method was found to increase it. It was difficult to purify the DNA sufficiently since sugars and phenols accumulate in leaves under heat stress (Wahid et al., 2007b).

In general, a self-mixed buffer combination was used but also two commercially available extracting kits tested did not achieve higher or purer DNA amounts. A probable reason for the low output might be possible pest infestations with viruses and/ or fungi. Since both possess incubation times of several days a sampling of already infected leaves can not be excluded.

For the AFLP analyes, the first used enzymes *Hin*dIII, as mentioned by Saliba-Colombani et al. (2000) and *Tru*I (a *Hin*dIII homolog), two rare-cutting restriction enzymes, did not restrict the tomato DNA properly. A reason might be the optimum working temperature of *Tru*I at 65 °C. The temperature difference between *Tru*I and the second frequent cutting enzyme *Mse*I was around 30 °C and therefore irreconcilable. Though the optimum working temperature of *Hin*dIII and *Mse*I was given the same by the producer, results obtained by *Hin*dIII restrictions were not reproducible although the DNA fragments on the agarose gels showed an expected length of around 50-200 bp. Different requirements to the buffers might be a reason for the reduced activity.

Diskussion

PCRs were conducted using different unspecific and specific primers in the reactions. Due to the improper fermentation by the restriction enzymes, the gel electrophoreses showed band patterns but the bands were not reproducible. In repetitions of the PCRs using the same restricted DNA samples and primer combinations, the band patterns on the gel were generally the same but they differed for the single genotypes. A band visible for a specific genotype in one experiment was missing the next time or vice versa. Since the visible bands were at positions comparable to bands of other genotypes, the error was not detected immediately. In further experiments, another enzyme combination was used according to Beraldi et al. (2004), Tam et al. (2005) and Truong (2007). Using another enzyme combination the PCR fragments obtained after DNA restriction with *Eco*RI and *Mse*I showed a high reproducibility. Since especially *Eco*RI is very sensitive and tends to 'staractivity' the usage of proper buffers is essential (Mülhardt, 2006). 'Staractivity' leads to unspecific restriction sites resulting to low adaptor ligation and/ or unspecific band patterns after PCR reactions.

Using *Eco*RI and *Mse*I no differences were found in repeatedly investigated genotypes. In first analyses, several polymorphisms could be detected and the approach showed promise for further experiments therefore.

After the detection of polymorphic markers a linkage analyses becomes possible and a genetic linkage map can be constructed. To obtain reliable results for pairwise calculations of map distances the markers numbers have to be high enough.

Several linkage maps with different kind of molecular markers are published (Eshed et al., 1994; Fulton et al., 2000; Fulton et al., 2002) providing a good base for further investigations. Due to indications on QTLs on the chromosomes 2 and 5 in the ILs influencing parthenocarpy special emphasis has to be placed on detecting markers on these chromosomes.

A highly saturated linkage map provides possibilities for further QTL analyses and it might be useful for the application in marker-assisted selections (MAS) in tomato breeding programs. Based on QTL analyses, fine mapping, and/ or *in-silico* evaluations of data bases candidate genes might be isolated providing the chance for enhancement of heat tolerance due to cloning approaches.

Since the negative influence of heat stress on some traits measured might refer to lack of Ca^{2+}, emphasis should be stressed on detection of plants with higher Ca^{2+} tolerance as reported by Starck et al. (1994). They conveyed differences between varieties in their capability of Ca^{2+} relocation between different parts of the plant under heat stress. Jiang et al. (2001) demonstrated that Ca^{2+} is required under heat stress for the maintenance of antioxidant activity. Kleinhenz et al. (2002) confirmed that Ca^{2+} is required for the mitigation of adverse effects of heat stress. With the application of Ca^{2+} prior to heat stress treatments they stimulated the activities of peroxidases needed for cell detoxification. Kolupaev et al. (2005) suspected an increased peroxidase activity as reason for enhanced heat tolerance. On chromosome 2, in the segment on which we assume genes related to parthenocarpy, a COS (conserved ortholog set) marker associated with peroxidases was mapped in tomato (sol genomics network, www.sgn.cornell.edu, 01.04.2008).

Furthermore, at least two loci containing alleles for male sterility are known to be located on the long arm of chromosome two (sol genomics network, www.sgn.cornell.edu, 01.04.2008). Since male sterility caused by damaged pollen leads to high percentages of parthenocarpic fruits markers mapped in this region might therefore be of high interest for MAS of plants tending to lower parthenocarpy.

On chromosome 5, in the small, nine centimorgan enclosing segment 2-4 which might carry alleles involved in reduced parthenocarpy, COS markers associated with a proline-rich protein and a glycoprotein were mapped.

Proline is known to accumulate in large quantities in response to environmental stresses (Kishor et al., 2005). Wahid et al. (2007a) presumed a high potential of proline to buffer cellular redox potentials under heat stress. Under high temperatures, fruit set in tomato plants was demonstrated to fail due to the disruption of proline transport during the narrow window of male reproductive development (Sato et al., 2006).

Glycoproteins, especially lectins, play an important role in the stimulation of pollen tube growth (Matveeva et al., 2007) among other things.

Therefore, IL analyses combined with database search might give valuable hints for putative candidate genes influencing the set of fertilized fruits under heat stress to develop markers for the identification plants better adapted to high temperatures.

5 Conclusions

The tomato is by now one of the most valuable and widely grown vegetable in the world. It is widely cultivated mainly in temperate climate zones. Under tropical and subtropical climates heat stress is a severe constriction for tomato crop production resulting in poor fruit set and consequent low yields.

The aim of this study was the evaluation of reasons causing reduced fruit set under heat stress in tomato and the evaluation of possibilities for the enhancement of tomato production considering genetic diversity and cultivation conditions.

Our experiments revealed genotypically different stress responses of heat tolerant genotypes in terms of reduced number of inflorescences, flower numbers, fruit numbers, fertilization capabilities, and the pollen amount as well as the pollen viability. With increasing heat sensitivity of the plants the negative influence of heat stress on the traits was stronger.

Since all traits except the number of inflorescences and the ability of fruit setting were influenced by environmental factors low heritability for these traits has to be assumed. The low heritability impedes the usage of these traits for breeding purposes and further research since results are difficult to compare. The traits have to be investigated under several different environmental conditions to enable reliable conclusions.

We were able to prove negative influences of high temperatures on several traits related to plant growth and fruit set often combined with high air humidity. Our results supported earlier reports of Peet et al. (1997) and Peet et al. (2003) that plants react even more heat sensitive under higher air humidity levels.

In general, the vegetative growth was given evidence of stronger impairment of generative growth compared to vegetative development which was in line with results of Kinet et al. (1997), Peet et al. (2003) and Sato et al. (2006). Therefore, in the current studies we focused on the pollen viability and the pollen release from anthers. Both traits are vitally important for sufficient fertilization needed for the production of high quality and marketable fruits.

The different mulch colors in the GHs with intend of passive cooling the inside temperatures affected the plant growth positively in all seasons. The color of the ground mulch was found very effective on traits related to fertilization. White ground mulch was able to increase the pollen viability and reduce parthenocarpy.

Though the pollen amount was less affected by the cooling methods compared to pollen viability it tended to be increased in GHs with black ground mulch/ NIR-transmissive roofs and consequent

Conclusions

higher inside temperatures. Since elevated temperatures beyond a threshold temperature had a negative influence on the pollen amount the cooling methods might be unsuitable. The threshold temperature could not be defined yet and seemed to be dependent on other factors such as RH or soil temperature. It has to be determined for every genotype and every environment separately therefore. Exceeding the threshold temperature in our result led to an absolute loss of pollen amount.

In most experiments the NIR-reflecting pigment on the roofs did not affect the pollen viability and pollen amount and furthermore increased percentage of cracked fruits. Therefore, the application of the NIR-shading paint can not be recommended.

Furthermore, the application of higher mesh sizes with higher protection capability against insect pest infestations did not influence the pollen amount and viability. Therefore, the applications of higher mesh sizes have to be investigated further on since the reason for the reduced fruit set in GHs with higher mesh sizes found in earlier experiments (Harmanto et al., 2006) could not be clearified.

The cyclic rhythm of the pollen amount, first reported in the present study, has to be validated in further experiments. As long as no possibility is known to discontinue the cycles to obtain regular pollen shed over time genotypes should be searched with cycles as short as possible to ensure pollination and fertilization.

Because tomato breeding in the last decades focused on few traits only, the genetic erosion was high (Cebolla-Cornejo et al., 2007). Nevertheless, we detected considerable genotypic variation within *Solanum lycopersicum* genotypes for heat stress.

The pollen amount and pollen viability seem to react with different intense in temperature ranges not preventing pollen development completely. This demonstrates the necessity for further investigations of physiological processes in pollen development impaired by high temperatures. Furthermore, the pollen amount as well as the pollen viability might be used as selection criteria in breeding for heat stress since both traits varied between genotypes and were positively correlated with fertilization.

The differences in response of introgression lines to heat stress provide a good base for further research and QTL mapping since inferior results refer directly to a known chromosomal segment.

Independent of molecular approaches, the progenies of the best performing genotypes of the segregating F_2 population represent a good source for breeding purposes. The plants performing superior under heat stress have high potential of enhancing the heat tolerance in their progenies when crossed.

Because F_2 plants from an intraspecific cross do not hold the disadvantage of crosses with wild species they are of high value for breeding processes. The genotypes outperforming their heat tolerant parent might provide a good base for the enhancement of heat tolerance in breeding programms.

6 References

Abdalla, A. A. and Verkerk, K., 1967. Growth, flowering and fruit-set of the tomato at high temperature. Netherlands Journal of Agricultural Science 16, 71-76.

Abdul-Baki, A. A. and Stommel, J. R., 1995. Pollen viability and fruit-set of tomato genotypes under optimum-temperature and high-temperature regimes. Hortscience 30, 115-117.

Adams, S. R., Cockshull, K. E., and Cave, C. R. J., 2001. Effect of temperature on the growth and development of tomato fruits. Annals of Botany 88, 869-877.

Aloni, B., Pharr, M., and Karni, L., 2001. The effect of high temperature and high atmospheric CO_2 on carbohydrate changes in bell pepper (*Capsicum annuum*) pollen in relation to its germination. Physiologia Plantarum 112, 505-512.

Asins, M. J., 2002. Present and future of quantitative trait locus analysis in plant breeding. Plant Breeding 121, 281-291.

Atanassova, B., 1979. Comparative studies on the morphogenesis of longistylic and normal tomato cultiavars and lines. Tomtato Genetics Cooperative Report 29, 20-93.

Atherton, J. G. and Harris, G. P., 1986. Flowering. In: Atherton and Rudich (Eds.), The tomato crop. A scientific basis for improvement. Chapman and Hall, New York.

Baker, J. T., Allen, L. H., and Boote, K. J., 1992. Response of rice to carbon-dioxide and temperature. Agricultural and Forest Meteorology 60, 153-166.

Balatero, C. H., Hautea, D. M., Hanson, P. M., and Narciso, J. O., 2002. Development of molecular markers for marker-assisted breeding for bacterial wilt resistance in tomato. The Philippine Agricultural Scientist 85, 170-181.

Banon, S., Fernandez, J. A., Franco, J. A., Torrecillasa, A., Alarcon, J. J., and Sanchez-Blanco, M. J., 2004. Effects of water stress and night temperature preconditioning on water relations and morphological and anatomical changes of Lotus creticus plants. Scientia Horticulturae 101, 333-342.

References

Beraldi, D., Picarella, M. E., Soressi, G. P., and Mazzucato, A., 2004. Fine mapping of the parthenocarpic fruit (pat) mutation in tomato. Theoretical and Applied Genetics 108, 209-216.

Breto, M. P., Asins, M. J., and Carbonell, E. A., 1993. Genetic variability in *Lycopersicon* species and their genetic relationships. Theoretical and Applied Genetics 86, 113-120.

Brewbaker, J. L. and Kwack, B. H., 1963. The essential role of calcium ion in pollen germination and pollen tube growth. American Journal of Botany 50, 859-865.

Burk, E. F., 1929. The role of pistil length in the development of forcing tomatoes. Proceedings of the American Society for Horticultural Science 26, 239-240.

Camejo, D., Rodriguez, P., Morales, A., Dell'Amico, J. M., Torrecillas, A., and Alarcon, J. J., 2005. High temperature effects on photosynthetic activity of two tomato cultivars with different heat susceptibility. Journal of Plant Physiology 162, 281-289.

Carbonell, E. A., Asins, M. J., Baselge, M., Balansard, E., and Gerig, T. M., 1993. Power studies in the estimation of genetic parameters and the localization of quantitative trait loci for backcross and doubled haploid populations. Theoretical and Applied Genetics 86, 411-416.

Cebolla Cornejo.J., Soler, S., and Nuez, F., 2007. Genetic erosion of traditional varieties of vegetable crops in Europe: tomato cultivation in Valencia (Spain) as a case study. International Journal of Plant Production 2, 113-128.

Charles, W. B. and Harris, R. E., 1972. Tomato fruit-set at high and low-temperatures. Canadian Journal of Plant Science 52, 497-506.

Collard, B. C. Y., Jahufer, M. Z. Z., Rouwer, J., and Pang, E. C., 2005. An introduction to markers, quantitative trait loci (QTL) mapping and marker assisted selection for crop improvement: The basic concepts. Euphytica 142, 169-196.

Costa, J. M. and Heuvelink, E., 2005. Introduction: The tomato crop and Industry. In: Heuvelink (Ed.), Tomatoes. CABI Publishing, Oxfordshire.

References

Dane, F., Hunter, A. G., and Chambliss, O. L., 1991. Fruit-set, pollen fertility, and combining ability of selected tomato genotypes under high-temperature field conditions. Journal of the American Society for Horticultural Science 116, 906-910.

de Koning, A. N. M., 1994. Development and dry matter distribution in glasshouse tomato: a quantitative approach. Dissertation, Wageningen Agricultural University, Wageningen.

Deutsch, F., 2004. Erzeugung und Transformation haploider regenerativer Kallusse bei *Populus nigra* (L.) hybrida. Dissertation, Universität Hamburg, Hamburg.

Deutsch, F., Kumlehn, J., Ziegenhagen, B., and Fladung, M., 2004. Stable haploid poplar callus lines from immature pollen culture. Physiologia Plantarum 120, 613-622.

Dickinson, H., 1995. Dry stimas, water and self-incompatibility in *Brassica*. Sexual Plant Reproduction 8, 1-10.

Dinar, M. and Stevens, M. A., 1983. The effect of temperature and carbon metabolism on sucrose uptake by detached tomato fruits. Annals of Botany 49, 477-483.

Dominguez, E., Cuartero, J., and Fernandez-Munoz, R., 2005. Breeding tomato for pollen tolerance to low temperatures by gametophytic selection. Euphytica 142, 253-263.

Ebrahim, M. K., Zingsheim, O., El Shourbagy, M. N., Moore, P. H., and Komor, E., 1998. Growth and sugar storage in sugarcane grown at temperatures below and above optimum. Journal of Plant Physiology 153, 593-602.

El-Ahmadi, A. B. and Stevens, M. A., 1979. Genetics of high temperature fruit-set in the tomato. Journal of the American Society for Horticultural Science 104, 691-696.

Ercan, N. and Akilli, M., 1996. Reasons for parthenocarpy and the effects of various hormone treatments on fruit set in pepino (*Solanum muricatum* Ait.). Scientia Horticulturae 66, 141-147.

Eshed, Y., Abu-Abied, M., Saranga, Y., and Zamir, D., 1992. *Lycopersicum esculentum* lines containing small overlapping introgressions from *L. pennellii*. Theoretical and Applied Genetics 83, 1027-1034.

References

Eshed, Y. and Zamir, D., 1994. A genomic library of *Lycopersicon pennellii* in *Lycopersicon esculentum* - A tool for fine mapping of genes. Euphytica 79, 175-179.

Feder, M. E. and Hofmann, G. E., 1999. Heat-shock proteins, molecular chaperones, and the stress response: Evolutionary and ecological physiology. Annual Review of Physiology 61, 243-282.

Feder, N. and Obrien, T. P., 1968. Plant microtechnique - some principles and new methods. American Journal of Botany 55, 123-125.

Fernandez-Munoz, R. and Cuartero, J., 1991. Effects of temperature and irradiance on stigma exsertion, ovule viability and embryo development in tomato. Journal of Horticultural Science 66, 395-401.

Firon, N., Shaked, R., Peet, M. M., Pharr, D. M., Zamski, E., Rosenfeld, K., Althan, L., and Pressman, E., 2006. Pollen grains of heat tolerant tomato cultivars retain higher carbohydrate concentration under heat stress conditions. Scientia Horticulturae 109, 212-217.

Foolad, M. R., 2005. Breeding for abiotic stress tolerance in tomato. In: Ashraf and Harris (Eds.), Abiotic Stresses. Food Products Press, New York.

Franklintong, V. E. and Franklin, F. C. H., 1992. Gametophytic self-incompatibility in *Papver rhoeas*. Sexual Plant Reproduction 5, 1-7.

Fulton, T. M., Bucheli, P., Voirol, E., Lopez, J., Petiard, V., and Tanksley, S. D., 2002. Quantitative trait loci (QTL) affecting sugars, organic acids and other biochemical properties possibly contributing to flavor, identified in four advanced backcross populations of tomato. Euphytica 127, 163-177.

Fulton, T. M., Grandillo, S., Beck-Bunn, T., Fridman, E., Frampton, A., Lopez, J., Petiard, V., Uhlig, J., Zamir, D., and Tanksley, S. D., 2000. Advanced backcross QTL analysis of a *Lycopersicon esculentum* x *Lycopersicon parviflorum* cross. Theoretical and Applied Genetics 100, 1025-1042.

References

Fulton, T. M., Nelson, J. C., and Tanksley, S. D., 1997. Introgression and DNA marker analysis of *Lycopersicon peruvianum*, a wild relative of the cultivated tomato, into *Lycopersicon esculentum*, followed through three successive backcross generations. Theoretical and Applied Genetics 95, 895-902.

García-Alonso, Y., González, A., Espi, E., Lopez, S. J., and Fontech, A., 2006. New cool plastic films for greenhouse covering in tropical and subtropical areas. Acta Horticulturae 719, 131-137.

Gerlach, D., 1984. Botanische Mikrotechnik (3. Ed.). Eine Einführung. Georg Thieme Verlag, Stuttgart.

Goldman, M. H. S., Goldberg, R. B., and Mariani, C., 1994. Female sterile tobacco plants are produced by stigma-specific cell ablation. Embo Journal 13, 2976-2984.

Hall, A. E., 1990. Breeding for heat tolerance- an approach on whole plant physiology. Hortscience 25, 17-19.

Harmanto, Tantau, H. J., and Salokhe, V. M., 2006. Microclimate and air exchange rates in greenhouses covered with different nets in the humid tropics. Dissertation, Leibniz Universität Hannover, Hannover.

Havaux, M., 1998. Carotenoids as membrane stabilizers in chloroplasts. Trends in Plant Science 3, 147-151.

Hedhly, A., Hormaza, J. I., and Herrero, M., 2005. Influence of genotype-temperature interaction on pollen performance. Journal of Evolutionary Biology 18, 1494-1502.

Herrero, M. and Hormaza, J. I., 1996. Pistil strategies controlling pollen tube growth. Sexual Plant Reproduction 9, 343-347.

Heslop-Harrison, J., Heslop-Harrison, Y., and Shivanna, K. R., 1984. The evaluation of pollen quality, and a further appraisal of the Fluorochromatic (FCR) Test Procedure. Theoretical and Applied Genetics 67, 367-375.

References

Holdaway-Clarke, T., Feijo, J. A., Hackett, G. R., Kunkel, J. G., and Hepler, P. K., 1997. Pollen tube growth and the intracellular cytosolic calcium gradient oscillate in phase while extracellular calcium influx is delayed. The Plant Cell 9, 1999-2010.

Hormaza, J. I. and Herrero, M., 1996. Dynamics of pollen tube growth under different competition regimes. Sexual Plant Reproduction 9, 153-160.

Howarth, C. J., 2005. Genetic improvements of tolerance to high temperature. In: Ashraf and Harris (Eds.), Abiotic stresses: plant resistance through breeding and molecular approaches. Howarth Press Inc., New York.

Iwahori, S., 1965. High temperature injuries in tomato. IV. Development of normal flower buds and morphological abnormalities of flower buds treated with high temperature. Journal of the Japanese Society for Horticultural Science 34, 33-41.

Jiang, Y. W. and Huang, B. R., 2001. Effects of calcium on antioxidant activities and water relations associated with heat tolerance in two cool-season grasses. Journal of Experimental Botany 52, 341-349.

Karim, M. S., Percival, G. C., and Dixon, G. R., 1997. Comparative composition of aerial and subterranean potato tubers (*Solanum tuberosum* L.). Journal of the Science of Food and Agriculture 75, 251-257.

Khatun, S. and Flowers, T. J., 1995. The estimation of pollen viability in rice. Journal of Experimental Botany 46, 151-154.

Khavari-Nejad, R. A., 1980. Growth of tomato plants in different oxygen concentrations. Photosynthetica 14, 326-336.

Kinet, J. M. and Peet, M. M., 1997. Tomato. In: Wien (Ed.), The physiology of vegetable crops. CAB International, Wallingford, UK.

Kishor, P. B. K., Sangam, S., Amrutha, R. N., Laxmi, P. S., Naidu, K. R., Rao, K. R. S. S., Rao, S., Reddy, K. J., Theriappan, P., and Sreenivasulu, N., 2005. Regulation of proline biosynthesis, degradation, uptake and transport in higher plants: Its implications in plant growth and abiotic stress tolerance. Current Science 88, 424-438.

References

Kleinhenz, M. D. and Palta, J. P., 2002. Root zone calcium modulates the response of potato plants to heat stress. Physiologia Plantarum 115, 111-118.

Kleinhenz, V., Katroschan, K., Schuett, F., and Stuetzel, H., 2006. Biomass accumulation and partitioning of tomato under protected cultivation in the humid tropics. European Journal of Horticultural Science 71, 173-182.

Kolbe, A., Tiessen, A., Schluepmann, H., Paul, M., Ulrich, S., and Geigenberger, P., 2005. Trehalose 6-phosphate regulates starch synthesis via posttranslational redox activation of ADP-glucose pyrophosphorylase. Proceedings of the National Academy of Sciences of the United States of America 102, 11118-11123.

Kolupaev, Y. E., Akinina, G. E., and Mokrousov, A. V., 2005. Induction of heat tolerance in wheat coleoptiles by calcium ions and its relation to oxidative stress. Russian Journal of Plant Physiology 52, 199-204.

Kuo, C. G., Chen, B. W., Chou, M. H., Tsai, C. L., and Tsay, T. S., 1979. Tomato fruit set at high temperatures. In: Cowell (Ed.), 1st Intl. Symp. Trop. Tomato. Asian Vegetable Research and Development Center, Shanhua, Taiwan, 95-109.

Kuo, C. G., Chen, H. M., Shen, B. J., and Chen, H. C., 1989. Relationship between hormonal levels in pistils and tomato frui-set in hot and cool seasons. In: Green, Griggs, and McLean (Eds.), Proceedings of the International Symposium on integrated management practices at Tainan, Taiwan. Asian Vegetable Research Development Center Publication, Shanhua, 138-149.

Kuo, C. G. and Tsai, C. T., 1984. Alternation by high temperature of auxin and gibberellin concentrations in the floral buds, flowers, and young fruits of tomato. Hortscience 19, 870-872.

Levy, A., Rabinowitch, H. D., and Kedar, N., 1978. Morphological and physiological characters affecting flower drop and fruit set of tomatoes at high-temperatures. Euphytica 27, 211-218.

Lohar, D. P. and Peat, W. E., 1998. Floral characteristics of heat-tolerant and heat-sensitive tomato (*Lycopersicon esculentum* Mill.) cultivars at high temperature. Scientia Horticulturae 73, 53-60.

References

Maestri, E., Klueva, N., Perrotta, C., Gulli, M., Nguyen, H. T., and Marmiroli, N., 2002. Molecular genetics of heat tolerance and heat shock proteins in cereals. Plant Molecular Biology 48, 667-681.

Matton, D. P., Nass, N., Clarke, A. E., and Newbigin, E., 1994. Self-incompatibility - how plants avoid illegitimate offspring. Proceedings of the National Academy of Sciences of the United States of America 91, 1992-1997.

Matveeva, N. P., Lazareva, E. A., Klyushnik, T. P., Zozulya, S. A., and Ermakov, I. P., 2007. Lectins of the *Nicotiana tabacum* pollen grain walls stimulating in vitro pollen germination. Russian Journal of Plant Physiology 54, 619-625.

Morales, D., Rodriguez, P., Nicolas, E., Torrecillas, A., and Sanchez-Blanco, M. J., 2003. High-temperature preconditioning and thermal shock imposition affects water relations, gas exchange and root hydraulic conductivity in tomato. Biologia Plantarum 47, 203-208.

Mülhardt, C., 2006. Der Experimentator: Molekularbiologie, Genomics (5. Ed.). Elsevier, Spektrum Akademischer Verlag, München.

Mutwiwa, U. N., 2007. Effects of different cooling methods on microclimate and plant growth in greenhouses in the tropics. Dissertation, Leibniz Universität Hannover, Hannover.

Nasrallah, J. B., Stein, J. C., Kandasamy, M. K., and Nasrallah, M. E., 1994. Signaling the arrest of pollen tube development in self-incompatible plants. Science 266, 1505-1508.

Paterson, A. H., Damon, S., Hewitt, J. D., Zamir, D., Rabinowitch, H. D., Lincoln, S. E., Lander, E. S., and Tanksley, S. D., 1991. Mendelian factors underlying quantitative traits in tomato - comparison across species, generations, and environments. Genetics 127, 181-197.

Peet, M., Sato, S., Clémente, C., and Pressman, E., 2003. Heat stress increases sensitivity of pollen, fruit and seed production in tomatoes (*Lycopersicon esculentum* Mill.) to non-optimal vapor pressure deficits. Acta Horticulturae 618, 209-215.

Peet, M. M. and Bartholemew, M., 1996. Effect of night temperature on pollen characteristics, growth, and fruit set in tomato. Journal of the American Society for Horticultural Science 121, 514-519.

References

Peet, M. M., Sato, S., and Gardner, R. G., 1998a. Comparing heat stress effects on male-fertile and male-sterile tomatoes. Plant, Cell and Environment 21, 225-231.

Peet, M. M. and Willits, D. H., 1997. Response of ovule development and post-pollen production processes in male-sterile tomatoes to chronic, sub-acute high temperature stress. Journal of Experimental Botany 48, 101-111.

Peet, M. M. and Willits, D. H., 1998b. The effect of night temperature on greenhouse grown tomato yields in warm climate. Agricultural and Forest Meteorology 92, 202-203.

Poulton, J. L., Koide, R. T., and Stephenson, A. G., 2001. Effects of mycorrhizal infection and soil phosphorus availability on *in vitro* and *in vivo* pollen performance in *Lycopersicon esculentum* (Solanaceae). American Journal of Botany 88, 1786-1793.

Prasad, P. V. V., Boote, K. J., and Allen, L. H., 2006. Adverse high temperature effects on pollen viability, seed-set, seed yield and harvest index of grain-sorghum [*Sorghum bicolor* (L.) Moench] are more severe at elevated carbon dioxide due to higher tissue temperatures. Agricultural and Forest Meteorology 139, 237-251.

Pressman, E., Peet, M. M., and Pharr, D. M., 2002. The effect of heat stress on tomato pollen characteristics is associated with changes in carbohydrate concentration in the developing anthers. Annals of Botany 90, 631-636.

Rick, C. M. and Dempsey, W. H., 1969. Position of stigma in relation to fruit setting of tomato. Botanical Gazette 130, 180-189.

Rijstenbil, J. W., 2005. UV- and salinity-induced oxidative effects in the marine diatom *Cylindrotheca closterium* during simulated emersion. Marine Biology 147, 1063-1073.

Rivero, R. M., Ruiz, J. M., Garcia, P. C., Lopez-Lefebre, L. R., Sanchez, E., and Romero, L., 2001. Resistance to cold and heat stress: accumulation of phenolic compounds in tomato and watermelon plants. Plant Science 160, 315-321.

Rodriguez-Riano, T. and Dafni, A., 2000. A new procedure to asses pollen viability. Sexual Plant Reproduction 12, 241-244.

References

Rudich, J., Zamski, E., and Regev, Y., 1977. Genotypic variation for sensitivity to high-temperature in tomato - pollination and fruit set. Botanical Gazette 138, 448-452.

Ruttencutter, G. E. and George, W. L., 1975. Genetics of stigma position. Tomato Genetics Cooperative Report, 20-21.

Sachs, L., 1993. Statistische Methoden : Planung und Auswertung (7. Ed.). Springer Verlag, Berlin.

Sage, R. F., 1994. Acclimation of photosynthesis to increasing atmospheric CO_2 - the gas-exchange perspective. Photosynthesis Research 39, 351-368.

Sairam, R. K. and Tyagi, A., 2004. Physiology and molecular biology of salinity stress tolerance in plants. Current Science 86, 407-421.

Saliba-Colombani, V., Causse, M., Gervais, L., and Philouze, J., 2000. Efficiency of RFLP, RAPD, and AFLP markers for the construction of an intraspecific map of the tomato genome. Genome 43, 29-40.

Sato, S., Kamiyama, M., Iwata, T., Makita, N., Furukawa, H., and Ikeda, H., 2006. Moderate increase of mean daily temperature adversely affects fruit set of *Lycopersicon esculentum* by disrupting specific physiological processes in male reproductive development. Annals of Botany 97, 731-738.

Sato, S., Peet, M. M., and Thomas, J. F., 2000. Physiological factors limit fruit set of tomato (*Lycopersicon esculentum* Mill.) under chronic, mild heat stress. Plant Cell 23, 719-726.

Sato, S., Peet, M. M., and Thomas, J. F., 2002. Determining critical pre- and post-anthesis periods and physiological processes in *Lycopersicon esculentum* Mill. exposed to moderately elevated temperatures. Journal of Experimental Botany 53, 1187-1195.

Sawhney, V. K., 1983. The role of temperature and its relationship with gibberellic acid in the development of floral organs of tomato (*Lycopersicon esculentum*). Canadian Journal of Botany-Revue Canadienne de Botanique 61, 1258-1265.

Schiott, M., Romanowsky, S. M., Baekgaard, L., Jakobsen, M. K., Palmgren, M. G., and Harper, J. F., 2004. A plant plasma membrane Ca^{2+} pump is required for normal pollen tube growth and fertilization. Proceedings of the National Academy of Sciences of the United States of America 101, 9502-9507.

Schmidt, N., 2007. Vergleich verschiedener Methoden zur Untersuchung der Pollenvitalität bei Tomate (*Solanum lycopersicum* L.). B.Sc. Thesis, Leibniz Universität Hannover, Hannover.

Schoeffl, F., Prandl, R., and Reindl, A., 1999. Molecular responses to heat stress. In: Shinozaki and Yamaguchi-Shinozaki (Eds.), Molecular Responses to cold, drought, heat and salt stress in higher plants. R.G. Landes Co., Austin, Texas.

Scholberg, J., Mcneal, B. L., Jones, J. W., Boote, K. J., Stanley, C. D., and Obreza, T. A., 2000. Growth and canopy characteristics of field-grown tomato. Agronomy Journal 92, 152-159.

Spooner, D. M., Peralta, I. E., and Knapp, S., 2005. Comparison of AFLPs with other markers for phylogenetic inference in wild tomatoes [*Solanum L.* section *Lycopersicon* (Mill.) Wettst.]. Taxon 54, 43-61.

Starck, Z., Siwiec, A., and Chotuj, D., 1994. Distribution of calcium in tomato plants in response to heat stress and plant growth regulators. Plant and Soil 167, 143-147.

Stone, P., 2001. The effects of heat stress on cereal yield and quality. In: Basra (Ed.), Crop responses and adaptation to temperature stress. Food Products Press, Binghamton.

Suliman-Pollatschek, S., Kashkush, K., Shats, H., Hillel, J., and Lavi, U., 2002. Generation and mapping of AFLP, SSRS and SNPs in *Lycopersicon esculentum*. Cellular & Molecular Biology Letters 7, 583-597.

Tam, S. M., Mhiri, C., Kerkveld, A. V. M., Pearce, S. R., and Grandbastien, M.-A., 2005. Comparative analyses of genetic diversities within tomato and pepper collections detected by retrotransposon-based SSAP, AFLP and SSR. Theoretical and Applied Genetics 110, 819-831.

Tanksley, S. D., 1993. Mapping polygenes. Annual Review of Genetics 27, 205-233.

References

Tanksley, S. D., Ganal, M. W., Prince, J. P., de Vicente, M. C., Bonierbale, M. W., Broun, P., Fulton, T. M., Giovannoni, J. J., Grandillo, S., and Martin, G. B., 1992. High density molecular linkage maps of the tomato and potato genomes. Genetics 132, 1141-1160.

Tognetti, R., Johnson, J. D., Michelozzi, M., and Raschi, A., 1998. Response of foliar metabolism in mature trees of *Quercus pubescens* and *Quercus ilex* to long-term elevated CO_2. Environmental and Experimental Botany 39, 233-245.

Truong, T. H. H., 2007. Characterisation and mapping of bacterial wilt (*Ralstonia solanacearum*) resistance in the tomato (*Solanum lycopersicum*) cultivar Hawaii 7996 and wild tomato germplasm. Dissertation, Leibniz Universität Hannover, Hannover.

Urban, F., 2006. Histologische Untersuchungen zu den Auswirkungen von Hitzestress auf die Blütenentwicklung bei Tomate. Diploma Thesis, Leibniz Universität Hannover, Hannover.

von Malek, B., Weber, W. E., and Debener, T., 2000. Identification of molecular markers linked to Rdr1, a gene conferring resistance to blackspot in roses. Theoretical and Applied Genetics 101, 977-983.

Vos, P., Hogers, R., Bleeker, M., Reijans, M., Vandelee, T., Hornes, M., Frijters, A., Pot, J., Peleman, J., Kuiper, M., and Zabeau, M., 1995. Aflp - A new technique for DNA-fingerprinting. Nucleic Acids Research 23, 4407-4414.

Wahid, A. and Close, T. J., 2007a. Expression of dehydrins under heat stress and their relationship with water relations of sugarcane leaves. Biologia Plantarum 51, 104-109.

Wahid, A., Gelani, S., Ashraf, M., and Foolad, M. R., 2007b. Heat tolerance in plants: An overview. Environmental and Experimental Botany 61, 199-223.

Warnock, S. J., 1991. Natural habitats of *Lycopersicon* Species. Hortscience 26, 466-471.

Wessel-Beaver, L. and Scott, J. W., 1992. Genetic variability of fruit-set, fruit weight, and yield in a tomato population grown in 2 high-temperature environments. Journal of the American Society for Horticultural Science 117, 867-870.

White, P. J. and Broadley, M. R., 2003. Calcium in plants. Annals of Botany 92, 487-511.

Young, L. W., Wilen, R. W., and Bonham-Smith, P. C., 2004. High temperature stress of *Brassica napus* during flowering reduces micro- and megagametophyte fertility, induces fruit abortion, and disrupts seed production. Journal of Experimental Botany 55, 485-495.

Zhang, J. F., Lu, Y. Z., and Yu, S. X., 2005. Cleaved AFLP (cAFLP), a modified amplified fragment length polymorphism analysis for cotton. Theoretical and Applied Genetics 111, 1385-1395.

Appendices

Plant nutrition

Table 50: Average nutrient composition of Flory® 2 mega (Euflor GmbH, München, Germany) containing nitrogen (N), phosphorus (P), potassium (K), and magnesium (Mg) applied in the climate chambers with 1.5 ‰ concentrations. The fertigation solution was applied with every irrigation.

Nutrient	Concentration [mM]
N	11.4
P	1.9
K	6.7
Mg	1.4

Table 51: Average nutrient composition of the fertigation solution containing nitrogen (N), phosphorus (P), potassium (K), calcium (Ca), magnesium (Mg), sulfur (S), sodium (Na), boron (B), iron (Fe), copper (Cu), manganese (Mn), molybdenum (Mo), and zinc (Zn) applied in GHs in Thailand. The nutrient solution was dripped in intervals according to the calculated solar radiation integral.

Nutrient	Concentration [mM]	Nutrient	Concentration [μM]
N	7.4	B	6.0
P	0.8	Fe	4.2
K	5.9	Cu	5.3
Ca	3.1	Mn	3.0
Mg	0.7	Mo	1.1
S	1.7	Zn	1.4
Na	1.8		

Appendices

DNA extraction according to Engel (2005)

Durchführung in 1,5ml Eppis (konischer Boden) = *spitz*

- 20-25mg frisches Blattmaterial
- 200µl Extraktionspuffer zugeben
- vermahlen mit einem Mikrostößel, der in eine handelübliche Bohrmaschine eingespannt wird. Diese wiederum steht in einem Stativ
- 200 µl Extraktionspuffer zugeben und die Probe kurz vortexen
- Proben im Wasserbad 30min. bei 65°C inkubieren. *(hier bereits den 2. Durchgang beginnen)*
- 500µl Chloroform zugeben, gut vortexen
- abzentrifugieren bei 18°C und 5min 13000 rpm
- Überstand in ein frisches Eppi pipettieren
- 600µl CTAB Precipitations Puffer zugeben, vorsichtig schwenken, 15min.bei RT stehen lassen, erneut schwenken.
- 5min. bei 13000rpm abzentrifugieren⇒ Pellet am Boden des Eppis wird sichtbar
- Überstand abgießen und verwerfen (Pellet haftet in der Regel so fest, dass ein abpipettieren nicht notwendig ist.
- 600µl TE high salt auf das Pellet pipettieren und schwenken (wenn sich das Pellet schlecht löst, Reaktionsgefäße für einige Minuten bei 60-65°C in den Wärmeschrank.)
- 750µl EtOH 100% (bei -20°C gelagert) zugeben und mischen.
- 5min bei 13000rpm abzentrifugieren, abgießen und das Pellet ca. 30 min. trocknen lassen
- Pellet in 30-250µl TE 01 lösen

Dauer der Extraktion inkl. Trocknen: ca. 3h

Pro Extraktionsgang werden 24 Eppis bearbeitet. Die Arbeitsschritte können so geschachtelt werden, dass man pro Tag auf bis zu 240 DNA Extraktionen / AK kommt.

4-5 Gänge laufen dann parallel

Routine: 8 Gänge pro Tag

Dispenser Pipetten wichtig

Pistill mit Dest-Wasser reinigen, bei Alkohol pappt DNA, mit Tuch trocken reiben

Table 52: Ingredients, concentrations, and amounts of the extraction buffer used for DNA extraction according the protocol Engel (2005)

Extraction buffer		
Ingredient	Concentration	Volume [g]
CTAB		10

Ingredient	Concentration	Volume [mll]
Tris/ HCl	1 M	50
NaCl	5 M	154
EDTA	0.5 M	20
Replenish to 1 l		
Add 1 ml 2-Mercaptoethanol after autoclaving		

CTAB= Cetyl trimethylammonium bromide, EDTA= Ethylenediamine tetraacetic acid, NaCl= sodium chloride, Tris/HCl= trishydroxymethylaminomethane/ hydrochloric acid

Table 53: Ingredients, concentrations, and amounts of the precipitation buffer used for DNA extraction according the protocol Engel (2005)

CTAB precipitation buffer		
Ingredient	Concentration	Volume [g]
CTAB		10
Ingredient	Concentration	Volume [ml]
Tris/ HCl	1 M	50
Replenish to 1 l		

CTAB= Cetyl trimethylammonium bromide, Tris/HCl= trishydroxymethylaminomethane/ hydrochloric acid

Table 54: Ingredients, concentrations, and amounts of the TE high salt suspension used for DNA extraction according the protocol Engel (2005)

TE high salt		
Ingredient	Concentration	Volume [ml]
Tris/ HCl	1 M	10
NaCl	5 M	200
EDTA	0.5 M	2
Replenish to 1 l		

EDTA= Ethylenediamine tetraacetic acid, NaCl= sodium chloride, Tris/ HCl= trishydroxymethylaminomethane/ hydrochloric acid

Table 55: Ingredients, concentrations, and amounts of the TE 01 salt suspension used for DNA extraction according the protocol Engel (2005)

TE 01		
Ingredient	Concentration, pH	Volume [ml]
Tris/ HCl	1 M, pH8	1
EDTA	0.5 M	0.2
Replenish to 1 l		

EDTA= Ethylenediamine tetraacetic acid, Tris/ HCl= trishydroxymethylaminomethane/ hydrochloric acid

Appendices

AFLP Protocol according Truong (2007)

1. Digestion

Digest 15 μl of DNA template (500ng) with 15μl digestion cocktail at 37^0C for overnight (stove) or 4 hours (heating block)

Cocktail	Concentration	Volume (μl/tube)
MiliQ H_2O		9.8
10X buffer 2 (BioLabs)		3.0
*Eco*RI (BioLabs)	20 U/μl	0.8
*Mse*I (BioLabs)	10 U/μl	1.2
10X BSA		0.2
Total		15.0

Take 5μl of digestion product will load on 1% agarose gel (marker: 5μl of 100pb ladder). Sample will smear about of 100-1000pb. Incubate digestion product at 70^0C for 15 minutes to stop enzyme reaction.

2. Ligation

Prepare adaptors

+ For *Eco*RI Adapter pair (final concentration of 5μM):
    - 25 μl of *Eco*RI Forward Adaptor (conc. 100μM)
    - 25 μl of *Eco*RI Reverse Adaptor (conc. 100μM)
    - 450 μl of TE
       Total: 500 μl

+ For *Mse*I Adapter pair (final concentration of 50μM):
    - 250 μl of *Mse*I Forward Adaptor (conc. 100μM)
    - 250 μl of *Mse*I Reverse Adaptor (conc. 100μM)
       Total: 500 μl

After mixing the adaptors, heat at 95^0C for 5 min to denature. Then allow cooling slowly in a Styrofoam box to renature completely. Store at -20^0C

Take 20μl of digestion ligase with 10μl of ligation cocktail and incubate overnight at 16^0C or over weekend at 12^0C. Unused portion of the digestion mixture may be stored at -20^0C

Cocktail	Concentration	Volume (μl/tube)
MiliQ H$_2$O		6.4
10X Ligase buffer (BioLabs)		1.0
EcoRI adaptor (BioLabs)	5 μM	1.0
MseI adaptor (BioLabs)	50 μM	1.0
T4 DNA ligase	400 U/μl	0.4
Total		10.0

Spin down ligation and dilute 5-10X (LD). Store sample -20^{0}C.

3. Pre-amplification (AFLP1 Program)

Cocktail	Concentration	Volume (μl/tube)
MiliQ H$_2$O		10.0
10X PCR buffer		2.0
E primer (E-A)	10 μM	0.6
M primer (M-C)	10 μM	0.6
dNTPs	2.5 mM	1.6
Tag polymerase		0.2
DNA (LD)		5.0
Total		20.0

Perform a 1:30 or 1:20 dilution depends on sample (PD). Both unused diluted and diluted reactions can be stored at -20^{0}C (Diluted reaction can be store at 4^{0}C if use daily). Take 10μl of DP to load on 1% agarose gel (marker: 5μl of 100pb ladder).

4. Amplification (AFLP2 Program)

Cocktail	Concentration	Volume (μl/tube)
MiliQ H_2O		9.1
10X PCR buffer		2.0
E primer* (E-ANN)	1uM	1.0
M primer (M-CNN)	5uM	1.2
dNTPs	2.5mM	1.6
Tag polymerase	5U/μl	0.1
DNA (LD)		5.0
Total		20.0

* for labeled primer. For unlabeled primer, concentration will be 5pmole/μl.

Add 10μl stop buffer to 20μl amplified product and denature at 94^0C for 5 minutes

Solutions and ingredients used for PCRs

Table 56: List of adaptors and primers used for AFLP analyses according protocol 2

Name	Enzyme	Type	Sequence (5'.........3') →
EcoAdapt1	*Eco*RI	adaptor	CTCGTAGACTGCGTACC
EcoAdapt2	*Eco*RI	adaptor	AATTGGTACGCAGTC TAC
MseAdapt1	*Mse*I	adaptor	GACGATGAGTCCTGAG
MseAdapt2	*Mse*I	adaptor	TACTCAGGACTCAT
Eco +0	*Eco*RI	primer + 0	GACTGCGTACCAATTC
Eco +C	*Eco*RI	primer + 1	GACTGCGTACCAATTC C
Eco +CGA*	*Eco*RI	primer + 3	GACTGCGTACCAATTC CGA
Mse +0	*Mse*I	primer + 0	GACGATGAGTCCTGAGTAA
Mse +CTC	*Mse*I	primer + 3	GACGATGAGTCCTGAGTAA CTC
Mse +CTG	*Mse*I	primer + 3	GACGATGAGTCCTGAGTAA CTG
Mse +CTT	*Mse*I	primer + 3	GACGATGAGTCCTGAGTAA CTT
Mse +CAT	*Mse*I	primer + 3	GACGATGAGTCCTGAGTAA CAT
Mse +CAC	*Mse*I	primer + 3	GACGATGAGTCCTGAGTAA CAC
Mse +CAA	*Mse*I	primer + 3	GACGATGAGTCCTGAGTAA CAA
Mse +CGA	*Mse*I	primer + 3	GACGATGAGTCCTGAGTAA CGA
Mse +CGG	*Mse*I	primer + 3	GACGATGAGTCCTGAGTAA CGG
Mse +CCC	*Mse*I	primer + 3	GACGATGAGTCCTGAGTAA CCC

Table 57: Ingredients and concentrations of the 10 x Williams's buffer for AFLP-PCR

Ingredients	Concentration [mM]
Tris/ HCl	100
KCl	50
$MgCl_2$	20
gelatin	0.01 %

KCl= potassium chloride, $MgCl_2$= magnesium chloride, Tris/ HCl= trishydroxymethylaminomethane/ hydrochloric acid

Buffer solutions and running conditions of the gel electrophoreses

Table 58: Ingredients and concentrations of Longrun TBE buffer

Ingredients	Concentration [M]
Tris	1.340
Boric acid	0.440
Na2EDTA	0.025

Na_2EDTA= sodium ethylenediamine tetraacetic acid, Tris= trishydroxymethylaminomethane

Table 59: Ingredients and concentrations of 10x TBE

Ingredients	Concentration [M]
Tris	0.90
Boric acid	0.90
EDTA (pH8)	0.02

EDTA= ethylenediamine tetraacetic acid, Tris= trishydroxymethylaminomethane

Table 60: Ingredients and concentrations of the loading buffer

Ingredients	Concentration
formamide	98.00 %
EDTA (pH8)	10 mM
pararosalinine	0.05 %

EDTA= ethylenediamine tetraacetic acid

Table 61: AFLP electrophoresis conditions of the DNA analyzer (Gene readir4200, LI-COR Biosciences GmbH, Bad Homburg, Germany)

Pre-run	Settings
Volts	1500 V
Current	35 A
Power	35 W
Time	00:20 min

Run	Settings
Volts	1500
Current	35
Power	35
Time	04:00
Temperature	45
Scan Speed	3- moderate

Results for the segregating F₂ population

Table 62: Average number of flowers (Flav Inf⁻¹), fruits (Frav Inf⁻¹), the percentages of seeded fruits (Frseed), fruit set (Frset), viable pollen (Poll) and pollen amount (Poll Fl⁻¹) measured in 174 genotypes of a segregating F2 population and its parental lines (heat sensitive [P₁], heat tolerant [P₂], the first filial generation [F₁]) grown under heat stress (32/ 28 °C) for 12 weeks. Different lower case or upper case letters within columns indicate significant differences between genotypes at significance levels of α< 0.05 or α< 0.1, respectively. (SNK-Test, flowers per inflorescence: n= 2064; fruits per inflorescence: n= 1750; seeded fruits: n= 1392; fruit set: n= 1059; percentage of fertilized fruits: n= 1045; viable pollen: n= 775; pollen amount: n= 2259).

genotype	Flav Inf⁻¹		Frav Inf⁻¹			Frseed Inf⁻¹			Frseed [%]			Frset [%]			Poll [%]			Poll Fl⁻¹		
	n	mean		n	mean		n	mean		n	mean		n	mean		n	mean		n	mean	
1	9	6.22	abcd	8	0.75	ab	8	0.75	b	4	100.00	a	4	38.84	ab	5	23.94	A	15	71,614.63	abcdef
2	10	7.70	abcd	7	3.00	ab	6	2.17	b	6	95.83	ab	7	39.58	ab	5	47.50	A	17	48,841.97	abcdefgij
3	14	4.93	abcd	13	1.38	ab	13	1.23	b	10	86.67	abc	10	34.27	ab	4	25.50	A	17	60,955.91	abcdefgij
4	8	7.88	abcd	8	2.50	ab	8	2.50	b	6	100.00	a	6	64.26	ab	3	42.49	A	9	65,069.50	abcdefgij
5	10	7.60	abcd	10	1.60	ab	10	1.60	b	8	100.00	a	8	30.45	ab	4	29.07	A	15	36,198.00	abcdefgij
6	10	6.10	abcd	9	1.56	ab	9	1.44	b	5	96.67	a	5	42.68	ab	.	.	.	12	17,656.42	abcdefgij
7	12	6.50	abcd	10	1.70	ab	10	1.30	b	7	84.52	abc	7	43.27	ab	5	35.74	A	16	35,429.78	abcdefgij
8	13	6.54	abcd	11	1.27	ab	11	1.18	b	7	85.71	abc	7	30.73	ab	6	27.21	A	16	33,613.38	abcdefgij
9	11	5.82	abcd	9	1.78	ab	9	1.67	b	7	85.71	abc	7	46.66	ab	7	47.57	A	12	54,648.50	abcdefgij
10	14	6.64	abcd	12	1.75	ab	12	1.42	b	9	72.22	abcd	9	43.26	ab	3	29.80	A	15	29,895.90	abcdefgij
11	12	6.58	abcd	12	1.33	ab	12	1.33	b	5	100.00	a	5	55.36	ab	7	34.25	A	13	62,644.27	abcdefgij
12	10	7.00	abcd	8	3.38	ab	8	0.88	b	8	15.77	abcd	8	61.18	ab	3	20.81	A	10	51,296.95	abcdefgij
13	7	7.00	abcd	7	0.71	ab	7	0.71	b	.	.	.	.	.	.	4	15.64	A	8	37,793.06	abcdefgij
14	12	5.00	abcd	10	1.90	ab	10	1.30	b	8	77.08	abcd	8	51.79	ab	.	.	.	14	26,830.43	abcdefgij
15	11	4.91	abcd	10	1.20	ab	10	1.10	b	6	83.33	abcd	6	64.48	ab	7	41.78	A	10	54,515.70	abcdefgij

genotype	Flav Inf^{-1}			Frav Inf^{-1}			Frseed Inf^{-1}			Frseed [%]			Frset [%]			Poll [%]			Poll Fl^{-1}		
	n	mean		n	mean		n	mean		n	mean		n	mean		n	mean		n	mean	
16	10	5.10	abcd	9	2.11	ab	9	1.89	b	6	87.50	abc	6	69.48	ab	7	29.50	A	14	75,279.04	abcde
17	12	8.17	abcd	11	1.18	ab	11	0.73	b	5	46.67	abcd	5	25.52	b	.	.	.	14	62,254.54	abcdefgij
18	9	5.56	abcd	9	1.33	ab	9	1.22	b	5	93.33	ab	5	43.97	ab	.	.	.	11	10,213.14	efghij
19	14	8.64	abcd	11	1.64	ab	11	1.55	b	7	92.86	ab	7	34.64	ab	8	32.67	A	19	37,607.00	abcdefgij
20	11	3.64	bcd	11	1.00	ab	11	0.45	b	6	29.17	abcd	6	57.41	ab	5	27.75	A	11	61,434.73	abcdefgij
21	14	6.79	abcd	12	2.83	ab	12	0.67	b	8	26.39	abcd	8	63.29	ab	.	.	.	12	7,239.67	fghij
22	11	5.82	abcd	10	1.60	ab	10	1.23	b	6	83.33	abcd	6	42.17	ab	4	27.69	A	13	50,985.62	abcdefgij
23	12	8.33	abcd	12	1.83	ab	12	1.83	b	7	100.00	a	7	34.65	ab	4	55.12	A	16	36,201.28	abcdefgij
24	12	7.50	abcd	11	1.73	ab	11	1.36	b	7	82.86	abcd	7	32.85	ab	8	25.43	A	14	59,832.68	abcdefgij
25	11	7.09	abcd	11	2.91	ab	11	1.27	b	8	50.06	abcd	8	50.85	ab	3	47.12	A	16	26,191.50	abcdefgij
26	13	5.00	abcd	10	1.40	ab	10	1.30	b	6	94.44	ab	6	42.71	ab	5	22.44	A	10	50,203.25	abcdefgij
27	13	6.38	abcd	13	2.54	ab	13	1.23	b	12	45.14	abcd	12	56.41	ab	8	30.41	A	15	52,031.33	abcdefgij
28	.	.	.	.	.	.	.	.	.	.	.	.	.	.	.	.	.	.	.	.	.
29	5	6.20	abcd	3	0.00	b	3	0.00	b	.	.	.	.	.	.	.	.	.	.	.	.
30	13	6.31	abcd	11	2.00	ab	11	1.00	b	7	45.24	abcd	7	56.02	ab	6	35.94	A	11	43,011.41	abcdefgij
31	11	6.91	abcd	9	2.11	ab	9	1.00	b	7	47.62	abcd	7	46.41	ab	.	.	.	10	14,453.30	cdefghij
32	1	3.00	cd	1	3.00	ab	.	.	.	.	.	.	.	.	.	3	65.26	A	.	.	.
33	14	5.43	abcd	14	1.43	ab	14	1.43	b	10	100.00	a	10	57.83	ab	6	47.35	A	13	70,120.19	abcdefgh
34	18	6.56	abcd	16	1.06	ab	16	0.25	b	7	26.19	abcd	7	41.02	ab	.	.	.	14	9,609.50	efghij
35	14	6.07	abcd	11	2.09	ab	11	1.36	b	8	75.00	abcd	8	52.83	ab	4	21.09	A	17	25,413.74	abcdefgij

genotype	Flav Inf^{-1}		Frav Inf^{-1}			Frseed Inf^{-1}			Frseed [%]			Frset [%]			Poll [%]			Poll Fl^{-1}		
	n	mean		n	mean		n	mean		n	mean		n	mean		n	mean		n	mean	
36	9	7.78	abcd	9	2.11	ab	9	0.33	b	5	26.86	abcd	5	48.51	ab	.	.	.	17	10,156.38	efghij
37	12	7.92	abcd	11	2.82	ab	11	2.55	b	9	84.26	abc	9	46.00	ab	7	53.81	A	14	50,558.04	abcdefgij
38	12	5.17	abcd	9	0.67	ab	9	0.11	b	4	25.00	abcd	4	43.13	ab	.	.	.	13	8,317.46	fghij
39	16	7.75	abcd	15	1.40	ab	15	1.13	b	11	72.73	abcd	11	27.67	ab	6	43.13	A	16	41,963.00	abcdefgij
40	13	6.31	abcd	11	2.27	ab	11	1.73	b	9	70.00	abcd	9	41.56	ab	6	28.50	A	14	49,073.68	abcdefgij
41	9	5.56	abcd	8	0.88	ab	8	0.63	b	4	75.00	abcd	4	30.56	ab	3	31.49	A	14	37,522.43	abcdefgij
42	13	4.92	abcd	11	1.82	ab	11	0.73	b	7	50.00	abcd	7	63.10	ab	4	26.50	A	15	38,145.87	abcdefgij
43	10	7.70	abcd	10	0.80	ab	10	0.80	b	7	100.00	a	7	15.03	b	4	57.80	A	14	37,098.25	abcdefgij
44	14	6.07	abcd	12	1.25	ab	12	1.25	b	9	100.00	a	9	39.60	ab	5	39.43	A	15	25,489.67	abcdefgij
45	12	5.33	abcd	8	2.13	ab	8	1.13	b	6	58.06	abcd	6	60.35	ab	4	11.30	A	13	36,634.73	abcdefgij
46	9	8.67	abcd	9	2.11	ab	9	1.33	b	6	67.22	abcd	6	34.54	ab	.	.	.	8	14,199.38	cdefghij
47	7	6.29	abcd	7	2.29	ab	7	0.29	b	6	11.67	abcd	6	66.69	ab	.	.	.	8	12,285.31	cdefghij
48	12	6.42	abcd	11	1.27	ab	11	0.91	b	7	80.95	abcd	7	52.22	ab	8	47.55	A	12	41,185.00	abcdefgij
49	12	5.00	abcd	9	1.00	ab	9	0.89	b	5	90.00	abc	5	46.21	ab	5	7.00	A	10	47,203.15	abcdefgij
50	11	7.73	abcd	8	2.00	ab	8	2.00	b	5	100.00	a	5	46.49	ab	11	47.40	A	15	65,020.87	abcdefgij
51	16	6.63	abcd	13	1.08	ab	13	1.08	b	8	100.00	a	8	24.23	b	6	34.18	A	14	82,444.21	a
52	11	7.27	abcd	10	3.00	ab	10	2.10	b	9	65.83	abcd	9	53.65	ab	7	34.14	A	15	66,208.37	abcdefgij
53	13	6.54	abcd	10	1.40	ab	10	1.40	b	5	100.00	a	5	46.67	ab	6	30.65	A	15	58,895.83	abcdefgij
54	4	7.75	abcd	4	0.00	b	4	0.00	b	.	.	.	.	.	.	.	.	.	.	.	.
55	11	7.64	abcd	9	2.44	ab	9	2.44	b	6	100.00	a	6	46.33	ab	7	58.91	A	14	80,279.07	ab

genotype	Flav Inf^{-1}		Frav Inf^{-1}			Frseed Inf^{-1}			Frseed [%]			Frset [%]			Poll [%]			Poll Fl^{-1}		
	n	mean		n	mean		n	mean		n	mean		n	mean		n	mean		n	mean	
56	12	5.67	abcd	11	2.00	ab	11	1.00	b	8	51.25	abcd	8	43.68	ab	5	41.21	A	14	60,301.46	abcdefgij
57	12	8.17	abcd	11	0.73	ab	11	0.64	b	3	93.33	ab	3	27.06	b	7	32.65	A	12	45,768.29	abcdefgij
58	6	6.00	abcd	.	.	.	.	.	.	.	.	.	.	.	.	3	25.92	A	.	.	.
59	11	8.00	abcd	10	1.00	ab	6	1.67	b	6	100.00	a	6	25.16	b	7	28.08	A	15	28,447.97	abcdefgij
60	13	6.69	abcd	13	1.77	ab	10	2.20	b	10	90.00	abc	10	39.83	ab	4	55.15	A	14	67,209.89	abcdefgij
61	17	6.35	abcd	14	0.71	ab	8	1.13	b	8	87.50	abc	8	20.59	b	7	33.52	A	15	69,041.73	abcdefgi
62	11	6.45	abcd	11	3.27	ab	8	1.13	b	8	27.62	abcd	8	57.66	ab	.	.	.	15	41,489.67	abcdefgij
63	11	7.30	abcd	9	2.89	ab	7	1.71	a	7	55.95	abcd	9	41.71	ab	6	22.94	A	16	47,382.88	abcdefgij
64	13	6.92	abcd	11	2.09	ab	8	1.88	b	8	60.42	abcd	8	49.83	ab	5	34.05	A	13	52,908.69	abcdefgij
65	10	9.00	abc	9	2.33	ab	6	3.50	b	6	100.00	a	6	31.97	ab	8	39.23	A	15	41,218.87	abcdefgij
66	11	8.00	abcd	11	1.55	ab	8	1.63	b	8	76.25	abcd	8	36.44	ab	7	42.51	A	14	62,678.61	abcdefgij
67	14	7.86	abcd	12	1.58	ab	8	1.63	b	8	70.83	abcd	8	39.84	ab	7	44.61	A	14	39,096.07	abcdefgij
68	11	7.55	abcd	9	1.78	ab	5	2.80	b	5	86.67	abc	5	35.83	ab	.	.	.	17	20,404.56	abcdefgij
69	10	7.50	abcd	7	1.14	ab	3	2.67	b	3	100.00	a	3	25.11	b	5	34.51	A	7	66,160.86	abcdefgij
70	15	7.53	abcd	14	1.64	ab	8	1.13	b	8	30.83	abcd	10	41.99	ab	7	43.24	A	13	51,009.69	abcdefgij
71	12	5.92	abcd	12	0.33	ab	4	1.00	b	4	100.00	a	4	18.73	b	4	48.97	A	6	12,994.92	cdefghij
72	13	9.85	ab	13	3.92	ab	7	0.86	b	7	23.92	abcd	9	55.46	ab	.	.	.	16	16,816.50	bcdefghij
73	8	5.13	abcd	7	1.29	ab	3	3.00	b	3	100.00	a	3	58.73	ab	6	44.62	A	12	35,286.50	abcdefgij
74	12	9.75	ab	11	2.82	ab	9	2.00	b	9	80.56	abcd	11	34.07	ab	5	28.51	A	13	39,903.92	abcdefgij
75	15	6.87	abcd	13	1.31	ab	8	1.88	b	8	84.38	abc	8	28.75	ab	8	28.61	A	14	38,259.04	abcdefgij

genotype	Flav Inf^{-1}			Frav Inf^{-1}			Frseed Inf^{-1}			Frseed [%]			Frset [%]			Poll [%]			Poll Fl^{-1}		
	n	mean		n	mean		n	mean		n	mean		n	mean		n	mean		n	mean	
76	17	6.29	abcd	16	0.75	ab	8	1.50	b	8	100.00	a	8	23.44	b	6	34.44	A	15	41,385.47	abcdefgij
77	11	4.64	abcd	10	1.40	ab	5	2.20	b	5	73.33	abcd	5	64.76	ab	6	41.53	A	14	44,888.46	abcdefgij
78	9	5.89	abcd	9	0.56	ab	3	1.33	b	3	66.67	abcd	3	28.59	ab	6	46.01	A	11	49,644.95	abcdefgij
79	14	7.64	abcd	12	1.08	ab	4	2.50	b	4	85.42	abc	4	40.76	ab	6	26.08	A	19	33,215.55	abcdefgij
80	15	6.73	abcd	14	1.86	ab	5	1.80	b	5	34.38	abcd	5	60.71	ab	3	21.49	A	16	26,054.75	abcdefgij
81	13	7.08	abcd	13	2.85	ab	9	0.67	b	9	23.28	abcd	10	57.01	ab	6	30.13	A	11	7,272.77	fghij
82	15	7.07	abcd	15	0.40	ab	5	1.00	b	5	80.00	abcd	5	18.21	b	5	26.47	A	15	7,781.37	fghij
83	14	7.21	abcd	13	1.69	ab	5	1.00	b	5	33.33	abcd	5	46.50	ab	4	15.83	A	13	27,211.65	abcdefgij
84	5	5.40	abcd	5	0.40	ab	1	0.00	b	.	.	.	.	.	.	3	18.09	A	7	17,678.71	abcdefgij
85	11	7.00	abcd	11	0.73	ab	3	0.67	b	3	16.67	abcd	3	31.84	ab	.	.	.	7	5,156.43	gihj
86	13	7.31	abcd	12	0.92	ab	5	2.00	b	5	95.00	ab	5	31.07	ab	6	24.03	A	12	64,661.50	abcdefgij
87	12	6.92	abcd	11	1.82	ab	8	2.00	b	8	91.67	abc	9	37.81	ab	10	18.04	A	17	44,421.03	abcdefgij
88	14	4.93	abcd	12	1.25	ab	12	0.83	b	7	73.81	abcd	7	42.94	ab	8	29.08	A	13	76,550.50	abcd
89	13	6.85	abcd	10	3.80	ab	10	1.50	b	7	41.90	abcd	7	60.12	ab	3	46.37	A	13	42,043.35	abcdefgij
90	2	7.00	abcd	.	.	.	.	.	.	.	.	.	.	.	.	.	.	.	.	.	.
91	15	6.07	abcd	12	1.08	ab	12	0.67	b	7	72.38	abcd	7	39.17	ab	4	56.97	A	16	25,420.03	abcdefgij
92	12	6.42	abcd	10	0.40	ab	9	0.22	b	.	.	.	3	17.86	b	8	43.32	A	15	31,281.33	abcdefgij
93	10	7.70	abcd	10	2.50	ab	8	2.63	b	8	85.42	abc	8	41.78	ab	5	33.04	A	15	35,468.83	abcdefgij
94	10	3.90	abcd	.	.	.	.	.	.	.	.	.	.	.	.	.	.	.	.	.	.
95	14	7.14	abcd	10	2.60	ab	10	1.70	b	8	63.54	abcd	8	48.66	ab	.	.	.	14	8,716.61	fghij

genotype	Flav Inf^{-1}			Frav Inf^{-1}			Frseed Inf^{-1}			Frseed [%]			Frset [%]			Poll [%]			Poll Fl^{-1}		
	n	mean		n	mean		n	mean		n	mean		n	mean		n	mean		n	mean	
96	18	6.61	abcd	16	1.69	ab	12	1.50	b	12	61.11	abcd	12	33.66	ab	5	33.96	A	17	40,542.32	abcdefgij
97	16	7.38	abcd	14	1.21	ab	14	0.35	b	7	29.05	abcd	7	42.12	ab	.	.	.	13	8,617.92	fghij
98	16	6.00	abcd	13	1.00	ab	5	2.00	b	5	69.33	abcd	5	44.17	ab	3	37.12	A	14	14,241.21	cdefghij
99	14	7.00	abcd	12	1.75	ab	12	0.92	b	7	60.71	abcd	7	36.92	ab	5	24.45	A	14	41,685.36	abcdefgij
100	11	4.91	abcd	8	1.63	ab	8	1.63	b	6	100.00	a	6	42.06	ab	6	38.43	A	13	63,641.88	abcdefgij
101	12	5.50	abcd	8	1.25	ab	4	0.75	b	4	41.67	abcd	4	42.54	ab	.	.	.	13	16,322.23	bcdefghij
102	15	7.00	abcd	12	3.17	ab	12	1.50	b	10	55.00	abcd	10	51.18	ab	6	35.28	A	12	39,192.79	abcdefgij
103	14	6.29	abcd	11	1.27	ab	7	0.86	b	7	45.24	abcd	7	47.76	ab	.	.	.	16	10,673.94	defghij
104	5	4.60	abcd	5	0.20	ab	5	0.20	b	.	.	.	.	.	.	.	.	.	.	.	.
105	13	6.46	abcd	12	2.33	ab	12	0.50	b	7	35.71	abcd	7	52.70	ab	4	46.81	A	15	10,750.10	cdefghij
106	13	8.38	abcd	12	0.42	ab	3	1.67	b	3	100.00	a	3	19.86	b	3	32.86	A	11	27,358.05	abcdefgij
107	14	8.00	abcd	12	0.92	ab	12	0.83	b	8	93.75	ab	8	21.62	b	6	52.03	A	12	55,091.25	abcdefgij
108	12	6.75	abcd	10	1.20	ab	3	4.00	b	3	100.00	a	3	63.13	ab	5	41.84	A	12	45,338.63	abcdefgij
109	17	6.18	abcd	14	1.29	ab	7	2.14	b	7	85.71	abc	7	35.91	ab	7	32.66	A	17	45,992.71	abcdefgij
110	12	6.67	abcd	10	2.40	ab	10	2.20	b	8	84.38	abc	8	42.56	ab	8	32.96	A	16	76,728.53	abc
111	16	6.94	abcd	14	1.93	ab	7	3.86	b	7	100.00	a	7	46.53	ab	7	45.42	A	16	43,125.03	abcdefgij
112	4	7.25	abcd	.	.	.	.	.	.	.	.	.	.	.	.	5	37.21	A	.	.	.
113	11	7.27	abcd	10	1.80	ab	7	2.29	b	7	90.00	abc	7	39.88	ab	8	32.85	A	15	67,614.63	abcdefgi
114	11	5.91	abcd	10	0.40	ab	10	0.40	b	4	100.00	a	4	21.88	b	.	.	.	13	33,401.58	abcdefgij
115	10	9.20	abc	9	1.44	ab	9	0.56	b	8	34.38	abcd	8	18.24	b	6	24.70	A	16	50,078.16	abcdefgij

genotype	Flav Inf^{-1}		Frav Inf^{-1}			Frseed Inf^{-1}			Frseed [%]			Frset [%]			Poll [%]			Poll Fl^{-1}		
	n	mean		n	mean		n	mean		n	mean		n	mean		n	mean		n	mean	
116	10	6.20	abcd	7	1.29	ab	4	2.25	b	4	100.00	a	4	31.44	ab	6	45.88	A	12	63,255.25	abcdefgij
117	12	7.25	abcd	9	2.11	ab	9	0.67	b	4	46.40	abcd	4	46.97	ab	.	.	.	15	17,698.00	abcdefgij
118	11	8.09	abcd	7	4.14	a	6	3.50	b	6	82.14	abcd	6	48.15	ab	5	25.85	A	15	50,656.30	abcdefgij
119	13	7.85	abcd	10	2.20	ab	10	1.80	b	7	83.33	abcd	7	41.03	ab	6	36.91	A	14	40,011.25	abcdefgij
120	14	7.14	abcd	12	1.83	ab	8	2.63	b	8	96.88	a	8	41.19	ab	8	33.87	A	16	48,466.84	abcdefgij
121	9	5.78	abcd	7	1.29	ab	7	0.86	b	4	54.17	abcd	4	53.82	ab	5	24.16	A	6	70,963.58	abcdefg
122	14	5.07	abcd	13	1.85	ab	13	0.54	b	9	31.85	abcd	9	62.17	ab	.	.	.	10	21,656.35	abcdefgij
123	12	6.25	abcd	11	1.27	ab	5	2.00	b	5	66.67	abcd	5	32.14	ab	6	25.25	A	15	29,187.63	abcdefgij
124	14	6.57	abcd	12	1.92	ab	12	1.17	b	9	57.04	abcd	9	45.37	ab	7	43.77	A	14	35,602.82	abcdefgij
125	13	5.54	abcd	12	1.50	ab	8	1.75	b	8	79.17	abcd	8	37.38	ab	12	33.79	A	16	51,689.50	abcdefgij
126	13	7.00	abcd	10	1.60	ab	10	1.40	b	5	85.00	abc	5	38.98	ab	5	24.57	A	17	26,571.76	abcdefgij
127	4	9.25	abc	1	4.00	a	1	4.00	b	.	.	.	.	.	.	4	56.14	A	13	34,459.23	abcdefgij
128	11	6.45	abcd	9	1.22	ab	5	2.20	b	5	100.00	a	5	29.22	ab	.	.	.	15	19,729.27	abcdefgij
129	19	6.32	abcd	16	0.56	ab	16	0.50	b	7	85.71	abc	7	23.26	b	8	42.98	A	15	37,375.10	abcdefgij
130	11	7.91	abcd	9	3.33	ab	8	2.63	b	8	72.71	abcd	8	49.09	ab	6	39.24	A	17	41,617.71	abcdefgij
131	14	7.71	abcd	12	1.08	ab	12	1.00	b	4	87.50	abc	4	43.13	ab	5	37.44	A	14	21,685.39	abcdefgij
132	6	9.50	abc	.	.	.	.	.	.	.	.	.	.	.	.	6	47.77	A	13	39,206.85	abcdefgij
133	18	5.50	abcd	15	1.47	ab	9	0.89	b	9	37.04	abcd	9	49.89	ab	5	28.76	A	10	14,781.35	bcdefghij
134	13	6.31	abcd	11	2.18	ab	10	0.80	b	8	38.69	abcd	9	58.38	ab	3	45.49	A	11	34,218.86	abcdefgij
135	14	4.93	abcd	12	0.00	b	.	.	.	.	.	.	.	.	.	.	.	.	17	1,103.09	j

genotype	Flav Inf⁻¹		Frav Inf⁻¹		Frseed Inf⁻¹		Frseed [%]		Frset [%]		Poll [%]			Poll Fl⁻¹		
	n	mean	n	mean	n	mean	n	mean	n	mean	n	mean		n	mean	
136	12	7.67 abcd	10	2.80 ab	8	2.38 b	8	67.50 abcd	8	46.05 ab	4	30.44	A	15	17,979.27	abcdefgij
137	12	4.50 abcd	10	0.80 ab	6	1.17 b	6	83.33 abcd	6	24.31 b	.	.	.	14	12,645.18	cdefghij
138	14	5.00 abcd	12	1.17 ab	12	1.08 b	7	97.14 a	7	41.58 ab	4	44.34	A	12	19,036.58	abcdefgij
139	18	6.67 abcd	15	0.87 ab	15	0.27 b	7	33.33 abcd	7	25.02 b	.	.	.	18	4,461.89	ihj
140	10	7.30 abcd	7	3.86 ab	7	3.57 b	6	94.44 ab	6	63.54 ab	10	46.86	A	15	51,968.83	abcdefgij
141	14	7.71 abcd	12	1.58 ab	12	1.00 b	7	63.33 abcd	7	42.41 ab	9	40.58	A	16	29,209.03	abcdefgij
142	12	7.17 abcd	10	2.00 ab	5	3.20 b	5	76.67 abcd	5	54.63 ab	6	38.37	A	16	30,664.16	abcdefgij
143	15	6.80 abcd	13	1.15 ab	13	1.00 b	9	83.33 abcd	9	25.82 b	5	30.34	A	17	28,180.26	abcdefgij
144	13	7.69 abcd	10	1.60 ab	10	1.40 b	6	93.33 ab	6	45.69 ab	8	31.41	A	16	54,189.50	abcdefgij
145	11	8.18 abcd	10	1.10 ab	10	1.00 b	6	94.44 ab	6	23.37 b	7	35.81	A	9	61,649.33	abcdefgij
146	9	6.78 abcd	9	1.78 ab	8	1.63 b	8	81.25 abcd	8	37.08 ab	6	47.11	A	13	65,781.31	abcdefgij
147	11	7.45 abcd	11	2.36 ab	8	2.00 b	8	71.88 abcd	8	39.29 ab	9	40.97	A	14	65,759.00	abcdefgij
148	10	6.70 abcd	7	3.14 ab	5	2.00 b	5	50.00 abcd	5	63.03 ab	5	48.25	A	14	40,323.75	abcdefgij
149	13	9.15 abc	9	1.89 ab	5	2.60 b	5	83.33 abcd	5	33.89 ab	7	47.27	A	13	67,788.50	abcdefgi
150	14	6.50 abcd	13	1.00 ab	8	1.63 b	8	100.00 a	8	25.86 b	3	54.30	A	14	35,424.18	abcdefgij
151	14	9.00 abc	14	0.86 ab	6	2.00 b	6	100.00 a	6	23.65 b	6	54.34	A	14	22,712.11	abcdefgij
152	12	7.67 abcd	10	2.50 ab	8	2.75 b	8	89.58 abc	8	37.70 ab	8	43.32	A	16	52,773.50	abcdefgij
153	12	6.25 abcd	10	1.30 ab	5	1.80 b	5	75.00 abcd	5	33.42 ab	5	46.44	A	14	25,948.75	abcdefgij
154	10	10.40 a	8	3.00 ab	5	0.80 b	5	10.00 bcd	5	45.00 ab	.	.	.	14	8,738.96	fghij
155	10	6.10 abcd	10	1.10 ab	6	1.83 b	6	100.00 a	6	30.19 ab	3	32.39	A	13	41,346.23	abcdefgij

genotype	Flav Inf^{-1}		Frav Inf^{-1}		Frseed Inf^{-1}		Frseed [%]		Frset [%]		Poll [%]		Poll Fl^{-1}	
	n	mean	n	mean	n	mean	n	mean	n	mean	n	mean	n	mean
156	7	6.00 abcd	.	.	.	.	.	.	.	.	.	.	5	16,125.10 bcdefghij
157	13	5.85 abcd	11	1.45 ab	8	1.75 b	8	92.71 ab	8	35.87 ab	5	27.68 A	15	36,427.17 abcdefgij
158	15	6.67 abcd	12	1.08 ab	6	1.83 b	6	79.17 abcd	6	34.56 ab	4	44.63 A	15	18,729.27 abcdefgij
159	8	6.75 abcd	8	0.88 ab	3	1.67 b	3	66.67 abcd	3	32.69 ab	7	35.13 A	12	55,143.33 abcdefgij
160	3	6.33 abcd	3	0.00 b	.	.	.	.	.	ab	.	.	.	.
161	8	7.88 abcd	8	1.13 ab	3	0.33 b	3	6.67 cd	3	37.14 ab	.	.	7	23,125.07 abcdefgij
162	15	6.73 abcd	11	1.64 ab	6	2.33 b	6	83.33 abcd	6	38.41 ab	4	43.88 A	16	26,640.72 abcdefgij
163	12	5.92 abcd	10	2.30 ab	8	1.50 b	8	50.00 abcd	8	52.68 ab	6	35.66 A	14	67,187.57 abcdefgij
164	10	6.00 abcd	7	1.71 ab	3	1.00 b	3	20.63 abcd	3	79.17 ab	.	.	10	3,437.70 ij
165	11	6.64 abcd	9	2.44 ab	7	2.29 b	7	65.48 abcd	7	55.86 ab	8	40.39 A	13	66,959.15 abcdefgij
166	18	6.28 abcd	13	1.46 ab	7	0.43 b	7	21.43 abcd	7	43.92 ab	.	.	14	8,426.50 fghij
167	8	6.13 abcd	6	1.67 ab	4	2.50 b	4	100.00 a	4	32.90 ab	5	50.64 A	16	47,841.84 abcdefgij
168	1	2.00 d	.	.	.	.	.	.	.	.	.	.	.	.
169	13	6.85 abcd	11	2.09 ab	9	1.78 b	9	62.22 abcd	9	39.21 ab	9	36.40 A	14	35,033.61 abcdefgij
170	2	8.00 abcd	.	.	.	.	.	.	.	.	.	.	.	.
171	12	7.25 abcd	11	2.00 ab	9	2.00 b	9	74.07 abcd	9	35.21 ab	8	30.99 A	16	75,615.28 abcde
172	12	7.67 abcd	10	1.60 ab	5	2.00 b	5	48.00 abcd	5	43.26 ab	4	55.69 A	16	34,648.47 abcdefgij
173	15	6.80 abcd	13	1.62 ab	8	2.13 b	8	89.58 abc	8	46.16 ab	3	49.58 A	16	20,468.88 abcdefgij
174	7	5.29 abcd	7	0.71 ab	2	2.50 b	.	.	.	ab	6	45.52 A	5	50,062.50 abcdefgij
F$_1$	23	6.42 abcd	19	0.85 ab	19	0.69 b	7	84.52 abc	7	38.97 ab	2	4.48 B	27	16,169.07 bcdefghij

genotype	Flav Inf^{-1}			Frav Inf^{-1}			Frseed Inf^{-1}			Frseed [%]			Frset [%]			Poll [%]			Poll Fl^{-1}		
	n	mean		n	mean		n	mean		n	mean		n	mean		n	mean		n	mean	
P$_1$	12	6.73	abcd	11	1.36	ab	11	1.35	b	5	100.00	a	5	37.84	ab	2	23.92	AB	11	33,480.18	abcdefgij
P$_2$	14	8.48	abcd	8	2.00	ab	8	2.00	b	7	100.00	a	7	49.99	ab	3	55.55	A	17	34,476.21	abcdefgij

Die VDM Verlagsservicegesellschaft sucht für wissenschaftliche Verlage abgeschlossene und herausragende

Dissertationen, Habilitationen, Diplomarbeiten, Master Theses, Magisterarbeiten usw.

für die kostenlose Publikation als Fachbuch.

Sie verfügen über eine Arbeit, die hohen inhaltlichen und formalen Ansprüchen genügt, und haben Interesse an einer honorarvergüteten Publikation?

Dann senden Sie bitte erste Informationen über sich und Ihre Arbeit per Email an *info@vdm-vsg.de*.

Sie erhalten kurzfristig unser Feedback!

VDM Verlagsservicegesellschaft mbH
Dudweiler Landstr. 99   Telefon +49 681 3720 174
D - 66123 Saarbrücken   Fax   +49 681 3720 1749

www.vdm-vsg.de

Die VDM Verlagsservicegesellschaft mbH vertritt

Printed by Books on Demand GmbH, Norderstedt / Germany